D1536540

MINORITIES IN SCIENCE
The Challenge for Change in Biomedicine

R
693
.M56
1977

MINORITIES IN SCIENCE

The Challenge for Change in Biomedicine

Edited by

VIJAYA L. MELNICK
Federal City College
Washington, D.C.

and

FRANKLIN D. HAMILTON
The University of Tennessee
and Oak Ridge National Laboratory
Oak Ridge, Tennessee

CENTER FOR BIOETHICS LIBRARY
KENNEDY INSTITUTE
GEORGETOWN UNIVERSITY

A PLENUM/ROSETTA EDITION

Library of Congress Cataloging in Publication Data

Main entry under title:

Minorities in science.

"A Plenum/Rosetta edition."
Includes bibliographies and index.
1. Minorities in medicine—United States. 2. Minorities in science—United States. 3. Medical education—United States. 4. Minorities—Education (Higher)—United States. I. Melnick, Vijaya L. II. Hamilton, Franklin D. [DNLM: 1. Minority groups. 2. Science. Q172 M666]
R693.M56 1977 507'.1 77-9465
ISBN 0-306-20027-9

Proceedings of a conference at the annual meeting
of the American Association for the Advancement of Science
held in Boston, Massachusetts, February, 1976

This project was supported by grant number 1R13GMMH23018-01
awarded by the National Institute of General Medical Science
and the National Cancer Institute, NIH, DHEW

A Plenum/Rosetta Edition
Published by Plenum Publishing Corporation
227 West 17th Street, New York, N.Y. 10011

First paperback printing 1977

© 1977 Plenum Press, New York
A Division of Plenum Publishing Corporation
227 West 17th Street, New York, N.Y. 10011

All rights reserved

No part of this book may be reproduced, stored in a retrieval system, or transmitted, in any form or by any means, electronic, mechanical, photocopying, microfilming, recording, or otherwise, without written permission from the Publisher

Printed in the United States of America

005116

This volume is dedicated to
George Mirron Willis

Chairpersons

Minorities in Science: Availability and Barriers That Affect Availability

Percy J. Russell
Department of Biology
University of California at San Diego
La Jolla, California

Problems of Minorities at Majority Institutions

Francine B. Essien
Associate Professor of Biology
Douglass College, Rutgers University
New Brunswick, New Jersey

Public Policy and Biomedical and Behavioral Training: Effective Development of Existing Potential

Vijaya L. Melnick
Associate Professor of Biology
Federal City College
Washington, D. C.

Cora B. Marrett
Associate Professor of Sociology
University of Wisconsin
Madison, Wisconsin

Financial Support for Minority Scientific Activities in Education and Research

Charles M. Goolsby *President*
Organization of Black Scientists
Washington, D.C.

Affirmative Action: Myth or Reality?

Franklin D. Hamilton
Associate Professor
The University of Tennessee–
Oak Ridge Graduate School of
Biomedical Sciences
Oak Ridge National Laboratory
Oak Ridge, Tennessee

Special Training Programs for Minority Students in Science: College Level

Edgar Smith
Provost
School of Medicine
University of Massachusetts
Worcester, Massachusetts

Special Training Programs for Minority Students in Science: Precollege Level

Cyrus J. Lawyer
Senior Program Associate
Institute for Services to Education
Washington, D.C.

Shirley Mahaley Malcom
Staff Associate
Office of Opportunities in Science
American Association for the Advancement of Science
Washington, D.C.

Contributors

James L. Angel, *Program Director, Medical College Admission Assessment Program, Association of American Medical Colleges, Washington, D.C.*

Alonzo C. Atencio, *Assistant Dean, University of New Mexico School of Medicine, Albuquerque, New Mexico*

Maxine Bleich, *Associate Program Director, Josiah Macy, Jr. Foundation, New York, New York*

Leroy T. Brown, *Assistant Vice Chancellor for Health Sciences, Health Science Opportunity Programs, University of Wisconsin, Madison, Wisconsin*

Theodore M. Brown, *Assistant Director, Center for Biomedical Education, City College of the City University of New York, New York, New York*

Jean V. Carew, *Graduate School of Education, Harvard University, Cambridge, Massachusetts*

Jewel Plummer Cobb, *Dean, Douglass College, Rutgers University, New Brunswick, New Jersey*

Therman E. Evans, *Executive Secretary, Health Manpower Development Corporation, Washington, D.C.*

Miles Mark Fisher, IV, *Executive Secretary, National Association for Equal Opportunity in Higher Education, Washington, D.C.*

Wayne Fortunato-Schwandt, *Office of Opportunities in Science, American Association for the Advancement of Science, Washington, D.C.*

J. N. Gayles, *Professor of Chemistry, Morehouse College, Atlanta, Georgia*

James C. Goodwin, *Assistant to the Vice-President, University of California at Berkeley, Berkeley, California*

Margaret S. Gordon, *Associate Director, Carnegie Council on Higher Education, Berkeley, California*

Zora J. Griffo, *Special Programs Officer, Office of the Director, National Institutes of Health, Bethesda, Maryland*

Franklin D. Hamilton, *Associate Professor, The University of Tennessee–Oak Ridge Graduate School of Biomedical Sciences, Oak Ridge National Laboratory, Oak Ridge, Tennessee*

Mary S. Harper, *Assistant Chief, Center for Minority Group Mental Health Programs, National Institutes of Mental Health, Rockville, Maryland*

Carl Hime, *Director, Office of Academic Services, Many Farms High School, Bureau of Indian Affairs, Many Farms, Arizona*

James M. Jay, *Professor of Biology, Wayne State University, Detroit, Michigan*

Sigfredo Maestas, *Dean of Academic Affairs, New Mexico Highlands University, Las Vegas, New Mexico*

Shirley Mahaley Malcom, *Staff Associate, Office of Opportunities in Science, American Association for the Advancement of Science, Washington, D.C.*

J. V. Martinez, *Division of Physical Research, Energy Research and Development Administration, Washington, D.C.*

James W. Mayo, *Deputy Division Director, Division of Science Education Resources Improvement, Directorate for Science Education, National Science Foundation, Washington, D.C. 20550*

Richard P. McGinnis, *Associate Professor of Chemistry, Tougaloo College, Tougaloo, Mississippi*

Bobbie N. McNairy, *Instructor of Mathematics, Prescott Junior High School, Baton Rouge, Louisiana*

Sidney A. McNairy, Jr., *Health Scientist Administrator, Minority Biomedical Support Program, Division of Research Resources, National Institutes of Health, Washington, D.C.*

Vijaya L. Melnick, *Associate Professor of Biology, Federal City College, Washington, D.C.*

Woodrow A. Myers, *Harvard University School of Medicine, Boston, Massachusetts*

Samuel M. Nabrit, *Executive Director, The Southern Fellowship Fund, Atlanta, Georgia*

Luis Nieves, *Program Director, Educational Testing Service, Princeton, New Jersey*

Stanford A. Roman, *Assistant Dean, Boston University School of Medicine, Boston, Massachusetts*

Edward R. Roybal, *Member, U.S. House of Representatives, Democrat–California*

Robert J. Schlegel, *Professor and Chairman of Pediatrics and Associate Dean for Public Policy, Charles Drew Postgraduate School of Medicine, Los Angeles, California*

William D. Wallace, *Director, Summer Programs, Harvard University, Cambridge, Massachusetts*

Robert A. Warren, *Associate Professor of Education, Texas A. & I. University of Kingsville, Kingsville, Texas*

Joseph W. Watson, *Provost, Third College, University of California at San Diego, La Jolla, California*

Foreword

Change is the essence of progress. We now stand at the crossroads of our civilization where change is essential in the conduct of our institutions, in the plans and models we project for the future, and in the very patterns of our thinking if we are to survive as "one nation under God . . . with liberty and justice for all."

Opportunity to participate and fulfill the responsibility of building the nation must be available to all citizens in a true republic. For the viability of governmental institutions, in a modern democratic nation-state, rests on the diversity of the genius of her citizens, and this enables the nation to accommodate herself better to changes of the times. But if the nation becomes impervious to change and resistant to modify its institutions to keep in pace with the times, then the nation will indeed be doomed to wither and perish. History is replete with examples of civilizations that have gone that course. It is therefore our responsibility to insure that our government institutions are kept receptive to change and reflective of the needs and concerns of her citizenry.

In America today, economic and social powers generally go to those who can claim a superior education and professional experience. As our society, and indeed the world, becomes increasingly dependent on science and technology, education in those fields becomes imperative to the power equation. Even though great changes have come about since the passage of the 1964 Civil Rights Legislation, those of us who are concerned with the status and destiny of the minorities and the disadvantaged of this society have to admit that we have yet a long way to go before the American dream becomes a reality for these citizens.

Knowledge is power, and as President Kennedy pointed out "Science is the most powerful means we have for the unification of knowledge; and a main obligation of its future must be to deal with problems which cut across boundaries, whether boundaries between the sciences, boundaries between nations or boundaries between man's scientific and human concerns."* It is my belief that science and the scientific endeavor will be enriched by the full participation of a broader group of people. For oftentimes, questions in science reflect the human experience of the investigator. Thus, the participation of citizens from all segments of society will ensure a more equitable and compassionate distribution of the fruits of science and application of technology.

I have been especially concerned with the status of health care in our country. If the access to good health care is indeed a right and not a privilege, then we need the involvement and the cooperation of a well-informed and competent group of citizens to help bring that about. This becomes especially important in the case of the minorities and the disadvantaged, for they, more than all others, bear the onus of an inequitable health-care system. We need more minority physicians and biomedical scientists, not only to improve the health-care status of their own communities but to enhance the health of this nation through the infusion of our citizen genius.

This volume contains the eloquently expressed concerns of a wide spectrum of men and women who came together in an effort to examine the present status of minorities in science and to seek and identify necessary strategies to improve the status in meeting the challenge of change in biomedicine. The editors of this book, Drs. Vijaya L. Melnick and Franklin D. Hamilton, are to be congratulated for their effort in bringing these papers together and making them available to the public through this volume. The papers rightly point to the current alarming underrepresentation of minorities in the biomedical professions, but they also speak of constructive efforts that can be designed and implemented to ameliorate the deficient situation. We shall do well to listen and take needed action to effect change that will sustain the viability of this potentially great society of ours.

ANDREW YOUNG
Ambassador Extraordinary and
Plenipotentiary, Permanent
Representative to the United Nations

*John F. Kennedy, 1963. Address to the National Academy of Sciences. *New York Times*, Oct. 23, 1963, p. 24.

Preface

By all accounts the conference on Minorities in Science (held as part of the AAAS Bicentennial Meeting in February 1976) was a success and, I trust, a forerunner of others. The conference featured thoughtful speakers and an interested, concerned audience. The report of these proceedings, the substance of this volume, should be read and studied with care by all members of the science community.

No social problem in science is more vexing than this issue of minority participation in the American science community. This community and the nation have made a commitment to ensure equality of opportunity to all persons. We have also made a commitment to accelerate minority entry into all walks of life. Once this has been said, it is abundantly clear that this commitment is far from fulfilled. The simple fact is that minorities are just beginning to enter the mainstream of American science. And, of course, their numbers are far from representative.

Yet progress—slow and frustrating, but progress still—is being made. In 1973–1974, for instance, 7.4% of medical-school enrollments were minority; in the same year, minorities accounted for 3% of the enrollments in natural-science fields in Ph.D.-granting institutions. Modest as these figures are, they represent a sizable increase in minority participation compared to a decade ago.

Despite the clear record of achievement in recent years, one detects disturbing signs that the rapid growth of minorities participating in graduate education has begun to level off. It may well be that in science, at least, we have reached the point where future gains will come only from far more fundamental changes throughout our entire educational system.

In brief, we not only must overcome any existing barriers at the graduate level, we must eliminate barriers all the way back to the preschool–education level. For instance, it is clear that unless a young person begins rigorous mathematical training at the junior-high level, the individual immediately begins to lose some options for a scientific career. At the University of California at San Diego, where I am, an entering freshman presenting only *minimum* eligibility requirements (and the University of California may, by law, accept only the top 12.5% of state high-school graduates) is severely handicapped in pursuing a major in science and in most of the social sciences. It should be noted, by the way, that the university is now making a strong effort to encourage minority secondary-school students to take algebra in junior high school and complete at least an additional two years of advanced mathematics in high school.

For my part, I believe the science community should consider a major curriculum (and teacher-training) effort at the elementary-school level. I realize the problems of the bureaucracy, the enormous scale, and all the other factors inhibiting change in such institutions. Nonetheless, we scientists did have a positive impact through curriculum reform at the high-school level, and it may be time to extend that impact to the elementary-school level.

I trust the following essays will encourage each reader to be more sensitive to the problem and to consider innovative ways to achieve full minority representation in every aspect of the American science community.

W. D. McElroy
President (1976)
American Association for the Advancement of Science

Chancellor
University of California—San Diego
La Jolla, California

Contents

Introduction

One of the most productive and fruitful exercises in our celebration of the American bicentennial was that of making an assessment of the progress achieved by the various segments of society that have helped build this country into a great nation. The last decades have produced profound social, economic, and technological changes in this country. These changes have greatly affected the lives of our people and have gone beyond regional boundaries to put this country in the forefront of the world community in terms of power, influence, and scientific eminence.

In the light of such achievements for the majority of American people, we can ask: How have the disadvantaged and the minorities fared during this development? For there is no denying that the measure of real progress can be no better than the progress made by the least fortunate, especially in the context of the democratic ideals upon which this nation is based.

It is the pondering of this question that brought together a group of individuals in an attempt to determine if some answers could be found. Since many of those in the initial discussion groups were biomedical scientists by avocation, a decision was made to examine the progress made by minority groups of this country in the field of biomedicine. We saw this task as a highly relevant one and as one within the limits of our competence to evaluate, as the advances made in this society result, in no small measure, from the many advances made in the fields of biology and medicine. After much deliberation, and through the combined efforts of many interested individuals, the format and program for a conference on "Minorities in Science" was assembled. The major objective of the conference was to measure not only the progress of minorities but also to critically assess those policies, programs, and projects that have been directed toward the amelioration of the obviously unequal and inadequate representation of minorities in the biomedical sciences, including the health professions. Such an evaluation had heretofore not been attempted on a national scale. The Office of Opportunities in Science of the American Association for the Advancement of Science agreed to sponsor the conference as part of their 1976 Annual Meeting in Boston, Massachusetts. The National Institutes of General Medical Sciences and the National Cancer Institute of the National Institutes of Health made a generous grant toward the financial support of the conference. Indeed, the conference was held as planned, and it brought together leading figures in public policy, educational programs, research institutes, private and public organizations, and other institutions that influence the participation and representation of minorities in the biomedical field. The conference also attracted the largest number of minority scientists, educators, students, policy makers, and representatives from the private and public sectors ever to be assembled in an AAAS annual meeting. These individuals had as their common concern the present and potential role of minorities in the American biomedical community.

This volume contains the proceedings of that conference. For the success of the conference and the cooperation of assembling the papers for the published proceedings, we are indebted to many. Foremost among these are Janet Brown, Head, Office of Opportunities in Science, AAAS; Arthur Herschman, Head, Meetings Office, AAAS; Alexander Hollaender, Associated Universities, Washington, D.C.; Elward Bynum, National Institute of General Medical Science; Nola Whitfield, National Cancer Institute; Thomas Malone, National Institutes of Health; Francine Essien, Rutgers University; James Mears and Elisabeth Zeutschel, Meetings Office, AAAS; Barbara Moorman and Robin Rogers, University of Tennessee, Knoxville; and Betty Ann Angel of the Health Professions Educational Service, Inc., Rockville, Md. For administrative and secretarial assistance, we are grateful to the staff of the Office of Opportunities in Science, including Wayne Fortunato-Schwandt, Shirley Mahaley Malcom, Jean L. Kaplan, Irene K. Papas, Debbie Baines, and Susan Posner.

Robert Ubell, Tom Lanigan, Harvey Graveline, and Georgia Prince of Plenum Press have been of immense help and we owe our special thanks to them. Last, but not least, our deep appreciation and thanks go to the many who made the conference a success and in particular to the session chairmen and chairwomen and the participants who authored the following papers. Without their efforts, this volume would not have been possible.

The papers that are presented in this volume are addressed to all who are concerned with true national progress; that progress can be achieved only by the full participation of all of the diverse segments of our society. The essays, therefore, speak to the public and to the professional, to the student and to the educator, to the expert and to the ordinary citizen, in an attempt to bring about an understanding and to provide the reason and the framework to produce strategies for change that will result in a truly equitable and just society. Achieving this, we believe, will mark the beginning of this nation's journey toward meeting the promises and the challenges of its third century.

Vijaya L. Melnick

Franklin D. Hamilton

I. Minorities in Science: Availability and Barriers That Affect Availability

1

Black Americans in the Sciences

JAMES M. JAY

Although Negroes were "invited" to America in 1619, it was not until 1837 that the first black American received the M.D. degree; this honor went to James McCune Smith, who earned said degree at the University of Glasgow (Scotland). The first black to earn the M.D. degree on home soil was David J. Peck, and his degree was conferred in 1847 by Rush Medical College. When Harvard University conferred the D.D.S. degree upon George F. Grand in 1870, he became the first black American dentist. Six years later, Edward A. Bouchet became the first black American to receive the Ph.D. in the sciences (physics). It should be noted that while Bouchet's doctorate was awarded only 10 years after the first doctorate by an American university, only one other black had earned a doctorate degree in the sciences by 1900. And now, 100 years after Bouchet's early quest of the science doctorate, there are well under 2,000 American-born blacks who hold this degree in the natural and physical sciences, and this number represents approximately 1% of all science doctorates in the United States pool. Except for some apparently significant improvements in the area of engineering, the dramatic increases in the numbers that are necessary to bring about a more respectable representation of blacks at this level appear not to be occurring.

On the pages that follow are some facts and figures relative to black scientists, some probable reasons for such low numbers, some prospects for the immediate future, and some of the tangible achievements of the group in question.

There appears to be no exact count of either black physicians, dentists, and engineers, or other physical or natural scientists at any degree level. In view of this fact, it is not surprising that estimates of numbers in all professions and at all degree levels cover rather wide

JAMES M. JAY • Professor of Biology, Wayne State University, Detroit, Michigan.

ranges. With respect to holders of science doctorates among blacks, the most reliable figure is 128 for the period 1876–1943. Greene (1946) uncovered 119 of these in his study, while an additional 9 not uncovered by Greene were uncovered by this author. Actual numbers become more difficult to come by in the ensuing years because of various factors, among which is a larger output per year. Among the estimates recently made for black holders of doctorates in the physical and natural sciences over the 1866–1975 period is a low of 1,200 (Jay, 1975). This is in contrast to much higher numbers projected by the Commission on Human Resources, National Academy of Sciences (1974) and by Wilburn (1974). The latter author estimated that a total of 924 science doctorates were awarded to blacks between the fiscal years 1958 and 1971 and that 200 were awarded in 1971 alone. Projections of this magnitude may indeed be more reflective of actual numbers produced, but until an actual count can be made, these high numbers should be treated as projections only. By the use of methods previously described (Jay, 1971), I have confirmed 907 subjects over the 1876–1975 period with an additional 248 on whom I have only partial information, for a total of 1,155.

The degree fields represented by the 907 confirmed subjects are essentially as previously noted (Jay, 1971), with the biomedical sciences accounting for 43.6%, followed by chemistry with 30.9%. The profile of American blacks who hold science doctorates continues to be essentially unchanged. They are southern-born, with a baccalaureate degree from a historically black institution and then a doctoral degree from a majority institution. Among majority institutions, only the University of Pittsburgh and the University of Illinois are represented among the top 20 institutions that award bachelor's degrees to blacks. A notable change in profile is the rather sharp increase in the number of terminal degrees from the once racially segregated universities in the South. Unfortunately, this increase has not come as an addition to the once relatively high output of the Big Ten universities. The leading doctorate sources for the 907 confirmed subjects are indicated in Table 1. The dominance of the Big Ten universities is still evident among the 28 leading sources.

The vast majority of the doctorate holders continue to be employed by academic institutions, with the historically black schools employing between 65 and 75% of this group. The most recent employment of 140 of the 148 recipients of doctoral degrees between 1970 and 1975 is

Table 1. Leading Doctorate Sources of the 907

Institution	No.	Institution	No.
Ohio State University	53	University of Pittsburgh	16
University of Iowa	44	University of California at Berkeley	15
University of Illinois	43	New York University	15
University of Michigan	43	Harvard University	13
Howard University	34	University of Kansas	13
Cornell University	33	Kansas State University	13
Wayne State University	33	Brown University	12
University of Chicago	31	University of Oklahoma	12
Michigan State University	31	University of Minnesota	11
Iowa State University	29	Indiana University	9
Penn. State University	28	Massachusetts Institute of Technology	9
University of Wisconsin	22	Syracuse University	9
Catholic University	21	University of Texas	9
University of Pennsylvania	20		
Purdue University	20		

presented in Table 2, with 99 employed by academic institutions and 69 of these being at predominantly black schools. This pattern of employment will probably continue in the years to come, since enrollment is projected to increase at black colleges. The late Professor Horace Mann Bond (1972) noted that "the doctorate among Negroes is in truth a degree tied to teaching" (p. 75).

While blacks represent about 12% of the American population, black academic doctorates in all fields represent 1% or less of all such degrees held by Americans (Jay, 1971). A substantial increase in the numbers and the percentage of blacks in the sciences is warranted, and steps to achieve that goal should be pursued immediately.

Table 2. Employment Patterns of 140 of the 148 Who Earned Doctorates over the Six-Year Period 1970–1975[a]

Type of Employment	No.	Percentage
Academic institutions	99	70.7
Historically black	69	69.7
Majority	30	30.3
Industry	17	12.1
Federal government	8	5.7
Postdoctoral study	9	6.4

[a]Seven were distributed among the following: medical practice, hospital laboratories, self-employed, medical student, U.S. military, and unemployed.

The production of Ph.D.'s from the black population in northern states is inadequate. This is a major contributing factor, I believe, to the low number of black Ph.D.'s. The relatively large proportion of black Ph.D.'s who are of southern origin should, in fact, be viewed as perhaps normal contributions to the pool. An example of the magnitude of the lack of production of black scientists by northern sources may be seen from the following. According to the U.S. Census for 1940, there were 12,865,518 black Americans. The three northern states of Illinois, Michigan, and Ohio represented 935,252, or 7.27% of the total. Of the 907 confirmed doctorates noted above, only 30 were born in these states. Three southern states (Georgia, North Carolina, and Texas) had 2,990,616 blacks in 1940, or 23.25% of the total, and produced 140 of the 907 Ph.D. recipients. From the statistics employing these six states, picked somewhat at random, the southern black doctorate source was approximately five times greater than the northern.

Some of the reasons for the paucity of black scientists with origins from northern states and cities may be the following:

1. High-school counselors who either "turn off" black students or counsel them out of the sciences.
2. Public-school systems with disciplinary problems that lead to the production of both uninspired and undereducated students.
3. Black newspapers that all too often see fit to give two-page coverage to a slain dope kingpin but no more than one inch, if any space at all, to the achievements of black scientists.

In this context, it is disappointing to note that only fifteen scientists were included in the Ebony Success Library volume on 1,000 successful blacks (1973). Of these fifteen, nine are college or university presidents and were perhaps included for that reason alone.

It is said by many that northern-born blacks do not seek the science fields because of a wider choice of other good opportunities. While there may be some truth to this, this choice is clearly not of the magnitude that it is generally thought to be. It is also said that more black students in the North go into the medical professions and that consequently fewer are left for the hard sciences. The majority of black physicians, dentists, and veterinarians still come from the South and not the North.

It is my estimate that any increase in the number of black Ph.D.'s in the immediate future will result from overall improvements and expansion in the science programs of black institutions and the resulting demands for more blacks on science faculties. The science programs at the black colleges have been improved over the past few years as a result of funding from programs such as the COSIP/MISIP of the National Science Foundation and the Minority Biomedical Support (MBS) program of the National Institutes of Health. (See James Mayo and Zora Griffo, this volume.) While the total amount of this funding has been miniscule by federal funding standards, it has gone to institutions that lacked so much that its impact has been significant. These institutions clearly represent the most likely source of black students for Ph.D. programs for the immediate future.

Although the statistics on blacks in science are rather grim, worthwhile achievements have been made nevertheless. For example, there have been national awards for research competition, election to high offices in scientific societies, membership in the National Academy of Sciences, and similar achievements. Persons in these ranks include Dr. David Blackwell, the first black member elected to the National Academy of Sciences, and the late Dr. Percy Julian, who was elected to the Academy in 1974. Dr. Lincoln Hawkins, at the Bell Laboratories, is a member of the National Academy of Engineering. There are at least two black members of the National Academy of Sciences' Institute of Medicine, including Dr. Geraldine Woods and Dr. Jewel Plummer Cobb. The president-elect of the American Chemical Society is a black chemist. A black has served as president of the American Public Health Association. There are two black scientists who are chancellors of major universities. One black scientist has been honored in a commemorative postal stamp. This is, of course, George Washington Carver. A black scientist is vice-president of a major industrial company. In spite of these gains, we need many more black scientists. I think that when the statement is made that there is an overproduction of scientists, it should be qualified to mean that the overproduction does not apply to minority scientists. Therefore, the case cannot be made too strongly for additional training of black scientists if present and future needs are to be met and equity achieved in the biomedical profession.

REFERENCES

Bond, H. M. *Black American scholars: A study of their beginnings.* Detroit: Balamp Publishers, 1972.

Commission on Human Resources. *Minority groups among United States doctorate level scientists, engineers, and scholars, 1973.* Washington, D. C.: National Academy of Sciences, 1974.

Ebony Success Library. 1,000 successful Blacks. Vol. 1. Nashville: Southwestern Company, 1973.

Greene, H. W. *Holders of doctorates among American Negroes.* Boston: Medor Publishing, 1946

Jay, J. M. *Negroes in science: Natural science doctorates, 1876–1969.* Detroit: Balamp Publishers, 1971.

Jay, J. M. Black Ph.D.'s. *Science,* 1975, *190,* 834–835.

Wilburn, A. Y. Careers in science and engineering for black Americans. *Science,* 1974, *184,* 1148–1154.

2

Spanish-Surnamed Americans in Science: Availability and Barriers

J. V. MARTINEZ

My intent is not to delve too deeply into history—scholars exist who can do that better; however, it is useful to keep Santayana's lesson in view and avoid repetition of mistakes. If we keep in proper perspective the nature of those attempts made to increase minority representation in science, perhaps eventually a meaningful effort can be found that will be acceptable to all. It is becoming evident that the legal statutes adopted to deal with the problems associated with the lack of comparative representation of minorities in all professions have not provided an increase of our numbers within the ranks. Before the legal statutes associated with equal opportunity were adopted, minorities were on the defensive—forever protecting what little they had and inching onward to gain what small additional respect was possible. It was not much and it was not in proportion to the increase in our population, but it was something. When the legal statutes were enacted, a new hope was seen. It then seemed worthwhile to initiate an offensive, now that the legal route was available. Just as quickly as the offensive moves were started, we saw the opposing team take extraordinary defensive moves that all but neutralized the newly enacted legal muscle. Now the federal agencies responsible for equal-opportunity and affirmative-action programs are so backlogged with litigation that for all intents and purposes the legal course is now barely visible. It is overgrown with suits and countersuits, affirmative action and affirmative discrimination. The lesson of this outcome is: one cannot legislate morality.

In this analysis of brief history it is suggested that we seek another

J. V. MARTINEZ • Division of Physical Research, Energy Research and Development Administration, Washington, D.C.

means to accomplish our goals while staying within the law. One must take heed of the guidance embodied in former President Johnson's words in his 1965 commencement address to Howard University: "We seek not just legal equality but human ability, not just equality as a right and a theory, but equality as a fact and equality as a result."

In a study published in 1974, Jackson and Cosca documented teachers' behavior toward Mexican-American students in the classroom. They visited 494 classrooms in 59 schools located in the Southwest and evaluated the frequency characteristics of teacher–pupil interactions. The random sample taken from 430 eligible schools included analyses of fourth, eighth, tenth, and twelfth grades. They found disparities in the amount of praise or encouragement given Chicano and Anglo students, disparities in the amount of acceptance or use of students' ideas by the teachers, and disparities in the amount of positive feedback by the teachers to the pupils. Quantifying the results, they showed that teachers in the study praised or encouraged Anglo pupils 35% more often than they did Chicano students, that teachers accepted or used ideas from Anglo pupils 40% more often than they did those of Chicanos, and that the teachers directed 21% more questions to Anglos than to Chicanos. These results were obtained in spite of the fact that the teachers knew their performance as instructors was under evaluation. It seems fairly evident that the attitude of teachers in this case contributed to the lack of academic achievement by the Chicano students. This situation is not new to most of us, but it is the first time that I am aware of that an objective report has been made on the subject.

Attitude is coupled with advocacy. There are very few of us, if any, that do not need some form of advocacy to assure success. Advocacy is an interpersonal relationship that assists all of us in personal growth. This relationship is transferable in that we in turn serve as advocates for others. I wish to emphasize that I am thinking of the need for an advocate to help eliminate barriers to success and want to make clear that I am not thinking of a benefactor. The former demands some measure of competence, while the latter does not. Because of the apparent difference in life styles, it takes a minority member a longer time to cultivate that rapport that allows a supervisor to serve as an advocate for him or her. We must attempt to delineate, if not establish and delineate, a group advocacy (as opposed to an individual advocacy) for minority groups in order to help solve the problem associated with the inadequate representation of minorities in science and engineering. We have been working long enough in groups composed almost exclusively of minorities and thus have been talking to one another. That action was necessary to develop our own perspectives. Now we must seek to integrate all those "minority committees" of the various professional societies to include nonminority members who can act to serve as advocates for the needs of minorities. Instead of these nonminorities' acting as individual advocates, they must realize that presenting views in a group (thus, group advocacy), or rather presenting them in a consistent manner, will be more effective. The time is at hand when an effective strategy must be developed cooperatively with the power structure. We must be prepared to articulate and defend our position. Negotiation may require making concessions counter to our perspectives, but a meaningful dialogue is now necessary if we are to direct our efforts toward future progress.

For the record, let me cite four criteria that I think form an acceptable basis for developing a working definition of a minority group:

1. The minority group is geographically identifiable. This means that members of such a group can be traced to residential areas exemplified by ghettos, barrios, and reservations.
2. The minority group is culturally and physically identifiable. Members of this group can be identified through physical appearance, language, and customs.
3. The minority group is usually assigned a low index of intellectual achievement. By any index that has been used, members of the minority group usually score among the lowest in any "standardized" test of intellectual achievement (largely a product of deficit research).
4. The minority group is found among the lower economic levels of society. Members of minority groups have relatively low income and as a whole have the least influential voice within the political process.

With respect to the Spanish-surnamed Americans, the principal components of this minority are the Mexican-Americans and the Puerto Ricans.

One drawback that seems unique to the Spanish-surnamed minority is that the relatively large number of immigrants from most Spanish-speaking countries are indistinguishable from representatives of the native Spanish-heritage minority. This should come as no surprise. Nevertheless, failure to take this into account has slowed the rate at which members of our minority group have been hired into key positions. These positions can provide an opportunity to accelerate the removal of obstacles to the societal evolution of a minority group. These positions are usually of such distinction that they can lead to an important contribution by permitting members of a minority to serve as role models in a highly visible manner. The responsibilities of such an office usually allow its occupant the mobility and security that serve to encourage other members of his or her group to aspire to similar positions. The economic contribution is of importance as well. Occupants of these positions provide for the influx of funds into the minority community.

Now let us turn to some interesting statistics. A manpower report originating with the National Science Board (1973) shows that in 1973 there were 185,000 Ph.D.'s in science and engineering in the United States, and about 1.2 million engineers. About one-half of 1% (0.5%!) of these Ph.D.'s were held by representatives of some one minority group. Within the Hispanic-American minority group, at present, it is possible to identify approximately 30 Ph.D.'s in chemistry out of about 25,000 in the country and about 12 Ph.D.'s in physics out of about 15,000. There exist about 10 Ph.D.'s in mathematics out of 14,000 and about 40 Ph.D.'s in the life sciences out of 50,000.

A report from the National Science Foundation (1975a) dated September 1975 concerning the technical makeup of the 1972 labor force shows that minority-group members represented about 4%, or 53,000, of the 1.3 million total number of scientists and engineers. The minority contribution to this pool was 29,000 engineers and 24,000 scientists. Of the 53,000 engineers and scientists, 60% were Oriental (Japanese, Chinese, Filipino, and Korean—according to the report), 30% were black, and 10% were other nonwhite races (American Indian, Hawaiian, etc.). The report goes on to show that in relation to the size of the minority groups, there were 134 Oriental engineers and 89 Oriental scientists per 10,000 population of Orientals and 3 black engineers and 4 black scientists per 10,000 blacks. For the "other nonwhite races," the ratios were less than 1 per 10,000. For comparison, the statistics for the total United States show that there were 46 engineers and 27 scientists per 10,000 population. The report also shows that during 1973, a total of 5,200 blacks, 2,500 Spanish-surnamed, 3,000

Orientals, and 200 American Indians were enrolled in engineering schools. This minority population was 10,900 of the 191,500 total student population. It is instructive to view these statistics in terms of the number of foreign-born scientists and engineers residing in the United States. Over the period of 1963–1973, according to a National Science Foundation (1975b) document, approximately 105,000 scientists and engineers immigrated to the United States. This means that on the average, 9,000 scientists and engineers immigrated to the United States each year between 1963 and 1973. Thus, a four-year influx of foreign scientists and engineers is comparable to the combined total number of all similarly trained professionals that are minority-group members born in the United States.

It is well known that minority students receive low verbal scores on Scholastic Aptitude Tests (SATs); there appears to be unquestionable evidence that the language factor is a strong obstacle to the attainment of good verbal scores. It is a particular handicap to a bilingual group in which neither language is spoken well.

Recently, I learned of an interesting episode regarding the application of standardized-test results used to admit women into an Ivy League university. It turned out that a report by the admissions office indicated that on the average, entering freshman women had substantially higher SAT scores. In a later published review of the distribution of marks, the administration questioned why the courses in home economics and teacher education displayed a proportionately higher number of the better marks within the university. It never occurred to these officials to note that discriminatory practices force women to enroll preferentially in these two departments. It appears that the officials expected these women to make average marks although they were selected on the basis of having higher SAT scores. In this case, the SAT scores were well correlated with accomplishments, but it is unfortunate that the fact had to be ascertained by the highlighting of an injustice.

One possible solution to the problem of an inadequate representation of minorities in science and engineering involves the development of educational institutions uniquely devoted to the problem. Such institutions would excel not only in the teaching of science but would also have a capacity within their faculties to conduct competitive research. The availability of the necessary facilities is taken for granted in this rather ideal model. Reflection on this idea shows that the *academic* elements of such a model now exist and are achievable. What is not readily available in any such existing institution is the additional commitment to deal in a special way with minorities—not that some have not tried, but rather that they have not had the necessary resources. For one thing, they do not have a sufficient number of minority scientists on the faculty. Whatever minority scientists they do have are not assigned positions that would allow them to serve as the role models required for attracting and retaining minority students.

In the model cited here, these minority faculty members would be able to contribute their knowledge to the solution of the particular social problems that minority students must address during the course of their development. These faculty members might serve as mentors, but more importantly, they would serve as advocates for minority students and help enlighten other faculty and administrators of the unique academic problems that minority students must overcome. Empathy and its practical working manifestations are the key components of this concept. In addition, such faculty and the proper administrative officials could communicate their concern and interest in a substantive manner to students in feeder schools that would serve as a student resource for the model institution of which I speak. For this reason, it would be expedient to locate such an institution, which I call a *resource center for minorities in science* (RCMS), in an area that would allow it to benefit from the geographic location of minority groups. For the Spanish-surnamed, this implies locating the institutions in the southwest United States, in Puerto Rico, and in some of the metropolitan centers of the United States.

The RCMS institution could be a significant asset to faculty and administrators at nearby smaller, non-Ph.D.-granting colleges. It could help them improve their research and teaching abilities and in this way improve the quality of their college faculty. The RCMS institution would have funds set aside to provide an adequate number of scholarships for deserving minority students. The institution in this model would not attempt to emulate categorically the traditional time-through-put mode of student training, but neither would it lower the standards of required performance. It would devote some of its resources to seeking improvements in the model. Finally, and most importantly, the institution would not be comprised exclusively of minority groups. There is a stage in the development of a minority professional at which the exclusion of nonminorities is definitely detrimental and must be avoided. One must operate under the realization that eventually both minority and nonminority peoples must work together, and what better way is there to learn to do so than if they both train together? However, in this model, the programs are designed to impact on minorities primarily, and that objective must remain paramount. It is also necessary to keep in mind that science is universal. There are not a minority science and a nonminority science. The universality of science must be used as a guidepost in our work toward the eventual elimination of these adjectives. Implementation of the concept of the RCMS does not mean one has to start from scratch. It is only necessary to select the most likely institution and provide the needed incentives to bring the model to reality.

REFERENCES

Jackson, G., and Cosca, C. The inequality of educational opportunity in the Southwest: An observational study of ethnically mixed classrooms. *American Education Research Journal,* 1974, *11,* 219–229.

National Science Board. *Science indicators—1972.* Washington, D. C.: National Science Foundation, 1973.

National Science Foundation. Racial minorities in the scientist and engineer population. *Science resource studies—Highlights, NSF 74-314.* Washington, D.C.: NSF, 1975a.

National Science Foundation. Selected employment and labor force characteristics of the redefined science and engineering population. *Science resource studies—Highlights, NSF 75-326.* Washington, D.C.: NSF, 1975b.

3

The New Medical College Admission Test: New Dimensions in Assessment

JAMES L. ANGEL

The three-to-one ratio of applicants to available first-year positions in medical school assures that two of every three candidates will be delayed or stopped by barriers erected by the admissions process. The admissions procedure is complex and demands the utmost in energy, careful thought, and comparison if all competitors are to be treated fairly.

One element of the barrier (or gateway for one-third) is the Medical College Admission Test (MCAT). The significance of the MCAT is highlighted by its successful history as a cognitive predictor of success in medical school and the consequent reliance admissions committees place on it as a measure of academic achievement in preparation for medical school.

The power of the MCAT as an admission variable is based on the weight given to the scores by committees in relation to grade-point averages and indicators of personal qualities obtained through interview, autobiographies, letters from advisers, and other relevant background information. If an admissions committee relies on cognitive data (MCAT scores, grade-point averages) as prime selector variables, then MCATs and school-performance records as barriers to admissions become highly significant factors. If personal qualities are studied carefully through available evaluation procedures and balanced in importance with tests and grades, then the importance of MCATs as barriers is diminished, and other selection procedures (assessment techniques) rise in importance.

JAMES L. ANGEL • Program Director, Medical College Admission Assessment Program, Association of American Medical Colleges, Washington, D.C.

I will address the MCAT as an admissions barrier for minorities, but I will diverge from simple historical perspective, since the instrument is undergoing major change and new tests with new objectives will be introduced in 1977. The current MCAT will see its last use in fall 1976.

It is not presumptuous to say that a major reason for revising the MCAT has to do with concerns about its importance in admissions and its ultimate effect on minorities. Social emphasis on the need for quality and the availability of health care is also a prime concern, and admissions officers are constantly criticized for "not selecting the right students," thus contributing to current health-care problems. I doubt if any of us would accept this simplistic criticism at face value, but it draws attention to the important role as gateway or barrier played by the admissions committee.

Selection is an imprecise science, but the Association of American Medical Colleges (AAMC) is committed to a major effort to improve both the cognitive and the noncognitive assessment instruments and techniques (such as interviews) used in admissions. I will discuss the MCAT today, but future reports out of the AAMC will reflect the heavy emphasis being placed on the assessment of personal qualities at admissions in relation to the qualities expected in the competent, concerned physician.

The issues discussed here are based on (1) the MCAT as an admission tool, especially as it relates to minority applicants, and (2) the efforts now under way at the AAMC to improve the overall quality of the assessment procedures used for selecting applicants for positions in medical school.

The current high status of the physician's career makes entrance into medical school a prized possession for many people. Statistics are well known on the number of candidates and the number of positions available in medical schools. About 45,000 students file over 300,000 applications for about 15,300 entering positions each year. This intense competition historically has worked to the disadvantage of minorities. A disproportionate number of minority members has never had opportunity to consider a physician's career as a possibility—a serious problem not within my purview to discuss now but within the objectives of this conference. I bypass this significant segment of history to discuss a few of the steps taken by the AAMC to attack the problems created by this inequity and to open up new avenues of opportunity that might that might lead to equal opportunity in education, in social mobility, in aspirations, and in acquisition.

In the late 1960s, when much of our society was being sensitized by the social turmoil going on in the country, the AAMC began to examine thoroughly its role in society, emerging with aggressive, positive responses to the various pressures and opportunities afforded by the climate of the times. An early step was the creation of the Office of Minority Affairs, headed by Dario Prieto, which has had wide involvement in and impact on basic issues plaguing our society in relationship to minority participation. The office has received constant generous support from the association and its constituent schools and has taken firm action in responding to pressing social concerns. The AAMC's constituents (the medical schools and the teaching hospitals) have made significant advances in recruiting and training members of minority groups to become full-fledged members of the medical profession. Many admissions committees have attempted to evaluate candidates in a way that will identify personal potentials that may be hidden by factors such as slightly lower grade-point averages or MCAT scores, cultural–ethnic inhibitions, or prejudices of the reviewing committee.

The Council of Deans of the Association of American Medical Colleges, through a report of an *ad hoc* committee convened in 1971, outlined significant steps it felt that the association and its constituents should take in examining the admissions process (AAMC, 1971). The pressure of inordinate numbers of applicants, concern over minority representation in the applicant pool and in admitted classes, and a general feeling of concern about the approach used by schools in evaluating candidates prompted the recommendations.

The AAMC's Division of Educational Measurement and Research, under Dr. James Erdmann, began an immediate study to determine the best ways to respond to this report. A survey with a group of constituents indicated strong support for studies to be conducted in admissions, with MCAT being an important part of this review. From this support, the Medical College Admissions Assessment Program (MCAAP) was organized. The MCAAP is a broad-based program of study of both the cognitive and the noncognitive factors of the assessment techniques used in admissions, relating such a study both to the selection procedures and to the problems of prediction encountered in any admissions techniques. Through the MCAAP, a task force was organized from constituents, including deans, admissions officers, students, faculty, and researchers in medical schools, as well as preprofessional advisers and representatives from related agencies such as the National Board of Medical Examiners and the American Medical Association Liaison Committee on Medical Education. This task force, under the chairmanship of Dr. Thomas Meikle of the Cornell School of Medicine, formulated recommendations that became the basis for executive-council action and guidelines for the MCAAP study (AAMC, 1973).

The MCAAP developmental effort has had the continued and extensive involvement of minorities and women. Based on the premise that the concerns and interests of any group will be better protected by full involvement of that group, the MCAAP has been an open study incorporating knowledgeable and informed people from its inception. The following review will indicate how this participation has occurred and will demonstrate the study's broad reliance on the work of members from many groups.

A full description of project activities is beyond the scope of this report. The three stages I will cover are (1) the planning of MCAAP projects; (2) the national task force; and (3) cognitive-test development. These should demonstrate the points to be made today regarding barriers to a career as a physician.

Initially, guidelines for the participation of constituents were prepared, and in cooperation with the AAMC's Office of Minority Affairs, a carefully conceived plan for the full involvement of minority members and women was designed. (The women's plan is not a part of this presentation.)

1. Through discussions with leaders from various minority sectors, mostly health related, an *ad hoc* committee for minority concerns in the MCAAP was formed immediately. Its purpose was to formulate a carefully conceived position paper that would take into account the significant issues related to minorities' access to medical school, with special emphasis on admissions screening techniques and their impact on minorities. Chaired by Dr. John Watson, of the School of Medicine of the University of California at San Francisco, this committee developed a paper that became an instrumental statement in the deliberations of the MCAAP national task force, which was organized shortly thereafter.

The position paper recommended (a) the development of a new, standardized, cognitive test; (b) research to find improved ways to assess the personal qualities of applicants; (c) full validation studies of assessment instruments; (d) future consideration of a criterion-referenced admissions exam; and (e) strengthened counseling and advising resources.

It was proposed that the science disciplines of biology, chemistry, and physics be presented separately in the new tests and that skills in handling of written materials be assessed.

For summary, the committee strongly urged a new test, related to the study and practice of medicine, that presents applicants with tasks that fairly assess their knowledge and skills. Assessment should be relevant and provide the means by which students demonstrate their preparation for medical school.

2. Through a series of regional conferences and related meetings, a national task force emerged. It provided a broad base for constituent input from medical and premedical levels. Besides the paper described above, position papers were presented by medical-school deans, admissions officers, undergraduate advisers, a committee studying personality assessment, and several individuals.

The task force was comprised of nearly one-third minority members, assuring not just adequate representation but a full cohort of voices to debate and discuss issues. Briefly, the recommendations of the task force cover the following:

A. An advisory committee within the AAMC will maintain a constant surveillance of admissions processes to assure ongoing research and to produce the best possible techniques and measuring instruments for the selection of candidates. This committee has been formed, chaired by Cheves Smythe of the University of Texas Medical School at Houston.
B. Validation of assessment instruments or techiques is an integral part of the MCAAP. This means that assessment instruments are to be evaluated constantly for their validity as selection devices. The evaluation of content, face, and concurrent validity is an essential first step, but a long-term effort will be launched to study admissions variables for their usefulness as predictors of success in medical school and practice.
C. The current MCAT (a decent predictor of success in medical basic sciences) is being used until new tests are ready.
D. New cognitive examinations are to replace the MCAT. To be introduced in 1977, they assess a student's ability to work with

information in written or quantitative form, and they test knowledge and problem-solving skills through separate exams in biology, chemistry, and physics. In the future, additional areas will be considered in which students may demonstrate proficiency, with behavioral sciences as one possibility.

E. The assessment of nonintellectual (noncognitive) factors of human behavior is the focus of a major proposal at the AAMC, which incorporates a research design for the project and suggests a course of action for conducting the necessary studies. The extent of this project will require outside resources to assure that it can be appropriately carried out.

F. A major information program has been launched, with intent to establish communication with admissions officers, applicants, advisers, and researchers. An initial manual in the cognitive-assessment project describes the new exams in detail. Later manuals will present test content and sample questions prior to test administration, the interpretation of scores, the use of the exams in counseling, significant research efforts, and other topics.

3. The third stage, cognitive test development, represents one of the most extensive efforts to date to attack the barriers facing students wishing to enter medical practice. The MCAAP Task Force Report clearly stated that every effort must be made to provide an assessment process for admissions that is equitable, that does not create conditions that require the "special" treatment of any person, and ensures the assessment of knowledge and skills as relevant to the task at hand. This effort means program objectives, specific goals, and programs designed to move us away from concepts such as disadvantaged groups, minorities, special groups, reverse discrimination, and other "phrases of inequity."

Through fully pledged support of the executive council of the AAMC, a request for a proposal was submitted to contractors in this country who could offer innovative ways to attack the problems at issue in cognitive assessment. Five major proposals of striking quality were received in response. Careful independent evaluations were conducted, including evaluations from minority members and women in medicine. The contract went to the American Institutes for Research (AIR) on two premises. First was their suggestion of an innovative procedure by which important prerequisites for the study and practice of medicine could be identified. Second was their proposal for the full participation and involvement of underrepresented groups. Their proposal, coupled with the AAMC's plan for full communication with users, should result in a significant advance in admissions testing.

The contributions of minorities to all project activities is a key factor in the development of the new exams. A careful construction of specifications and questions utilizing the extensive involvement of minority-group members in each phase was one significant step. A brief discussion of that process follows.

The content upon which any test is based becomes the determinant for content validity. Validation procedures usually involve a set of judgments by informed people who determine the important knowledge and skills that should be assessed as prerequisites to study. The proposal from the AIR expanded on this concept considerably.

Comprehensive outlines of science knowledge were developed in biology, chemistry, and physics. Skills considered important in the study and practice of medicine were identified and listed. Types of material from which most information is gleaned were identified and entered in the outline. A survey booklet was then organized for all of this material, and invitations were extended to over 300 persons from medical education and practice to serve as evaluators. Approximately 160 actually participated in the ratings. Their task was to rate knowledge topics, skills, and source materials on the basis of their importance as prerequisites to the study of medicine and to rate their choices a second time, this time in terms of their importance as prerequisites to the practice of medicine.

Over one-tenth of the raters on this national survey panel were members of minorities. As a result, the numbers were nearly representative of the general population but markedly overrepresented in relationship to the numbers of minority members in medical education or practice. This ratio was deliberate, however, since it was more important to get an adequate population for reliable ratings than it was to have a representative population.

In brief summary, the evidence shows broad agreement among both nonminorities and minorities as to important prerequisites for the study and practice of medicine. A full report on this study will be forthcoming within the next year, but the importance of this preliminary information rests in the identification of the important prerequisites and in the level of agreement found among the clusters of raters. This gives evidence that there may be a certain basic level of scientific knowledge and certain types of skills required for the study and practice of medicine, regardless of background.

With this information, the AIR has proceeded to develop a set of specifications for the new admissions examination, basing it entirely on the knowledge and skills judged by the raters as important in medical education or in practice. *Nothing else will end up in the test.* Topics were eliminated if they did not obtain adequate support from the panelists. Thus, there should be no irrelevant material in any of the six tests.

Several other steps have been followed in the preparation of the new exams to provide an equitable assessment tool. One has been the wide use of minorities as item writers. Another has been an extensive review of every test item by minorities and women, looking both for cultural and for sexual bias. Third, and a most significant feature, has been the tryout of items with several hundred students, especially minority students, from undergraduate universities in many parts of the country. Students have had the opportunity to work the items, to evaluate them in terms of bias or other standards of quality, and to provide open critiques with other students and local university faculty members. On the basis of these reviews, the AIR has been able to remove items judged by students to be discriminatory, of poor quality, tricky, or otherwise inappropriate.

Careful studies are done on all data to uncover indications of any problems that might arise with the exams. Research is constant to assure early adjustment of any inequities that might appear. Basically, the procedure in this project is the most extensive ever undertaken to develop a standardized instrument for admissions assessment that will be relevant, technically sound, and equitable, to the extent humanly possible, for every person wanting to obtain a coveted place in medical school.

Anxiety for many students reaches inordinate levels as they seek admission to medical school. Part of the problem is posed by admissions testing, which is always a point of apprehension and a puzzle to many. In the case of the MCAT, students have not been aware of content other than in the general sense. Furthermore, they have been misdirected by commercial publications that do not always give appropriate emphasis when describing the tests.

Strong support has been given by admissions officers and the AAMC staff to the concept of providing complete information to students about the content of the exams. The AIR has based the entire test-development process on the assumption that students will be told the topics to be covered by the test, the types of skills to be assessed, and the nature of the materials from which the test items are to be drawn. Consequently, a manual is being introduced presenting a complete outline of topics in scientific knowledge that students should know, a description and definition of the skills to be assessed, and a discussion of the types of source materials to be incorporated. Sample items will be included in this manual to acquaint students with the types of questions on the New MCAT and to give the realistic work experiences that should be of most help in their preparation. Frankly, it is expected that no student will need to go to a commercial organization to obtain preparatory information for the new admissions testing program. The AAMC feels a deep obligation to students, advisers, and admissions officers to furnish the essential ingredients for appropriate use of the new exams.

The publication of this information for students will be followed in the winter of 1976–1977 by a national workshop for admissions officers to introduce them to the test, its scoring procedures, and its potential for use in admissions. Premedical advisers will have access to all materials, including instructions for the use of the examinations in counseling and advising. Technical manuals will be furnished for researchers and others interested in the technical quality of the tests and their adherence to professional standards.

Medical schools and individual researchers from the United States and Canada, along with interested parties from other countries, will be invited to take part in validation studies and other research efforts in the new program. Evaluation of the performance of students and of physicians is an important part of the establishment of criteria against which admissions variables are to be studied. Linking common behaviors along the medical-education and practice continuum, and extending them downward to the assessment of applicants, is an important task yet to be completed. The AAMC is currently conducting a major research study that should contribute to this effort, a longitudinal study originally started with medical students of the entering class of 1956; 28 schools are involved in this study, along with a cohort of some 2,500 physicians who graduated from those schools in 1960. This cohort is participating in a major study linking the various admissions and performance variables of the 1956–1960 period with current experiences in their practices. This study promises to bring many benefits to the validation research of the current project.

The MCAAP task force had proposed that a new longitudinal study be started with our current students. We believe that the introduction of the new cognitive test as well as the plans for the improved assessment of the personal qualities of candidates should be a logical starting point for a new study. Thus, the first elements of a design for a comprehensive longitudinal study are being considered; this study should result in a continuing, systematic effort to improve the quality of candidate selection.

This presentation does not make the assumption that the inequities of admissions to medical school are all going to disappear with the advent of better assessment procedures. We take into account that the solution of inequities is a major social and personal responsibility. We contend, though, that the effect of overemphasis on cognitive criteria for admissions has been harmful to many candidates, and especially so for minorities. Improved cognitive instruments may assure more equitable measures of candidates, but they cannot remove the tendency of certain admission committees to place too much reliance on cognitive information.

Most admissions committees work constantly to establish better procedures for assessing the personal qualities of candidates, but some observers see a need to take more risks in relationship to the committees' judgments about those qualities. Such risks may be the only way in which selection criteria can be expanded to offset the detrimental effect of overreliance on grades and test scores.

Our predictors, whether they be cognitive or noncognitive in focus, should be looking for the unusual indications of potential—the kind of potential that does not always lend itself to easy decision making such as can be done with ranking in test scores. We submit that the profile for persons admitted to medical school should some day be parallel to the profile for the physicians needed to deliver quality health care in this country.

Suspicions are widespread that our recruitment and training system has not always produced responsive, ethical, competent physicians. Evidence shows that only in the past decade or so have there been expanded, diligent efforts to remove barriers to medical school and practice for women and minorities. The fact that nearly 50% of the minority applicants are admitted to medical school in contrast to a one-of-three rate of acceptance for all applicants gives concrete evidence of improved attention to the problem.

Further work needs to be done in the recruitment of minority members for our applicant pool. As recently as 1972, studies showed that only 4.7% of black college freshmen were planning professional medical careers (Johnson, Smith, and Parnoff, 1975). More recent evidence shows that the applicant pool of minority members seeking entrance to medical school has leveled off, and enrollments show a slight decline. These are problems that go beyond the MCAT. We would hope, however, that the broad involvement of women and minorities in the planning and construction of a new admission test for medical school will be one significant step in removing barriers for persons who, first of all, have been denied access and, second, are most needed in order to strengthen the country's health-care delivery system significantly.

REFERENCES

Association of American Medical Colleges. Final Report of the AAMC National Task Force with Recommendations for the Medical College Admissions Assessment Program, 1973.

Association of American Medical Colleges, Council of Deans. Ad hoc committee established to examine admissions problems, 1971.

Johnson, D. G., Smith, V. C., Jr., and Tarnoff, S. L. Recruitment and progress of minority medical school entrants 1970–72. *Journal of Medical Education,* 1975, *50,* 746.

4

The Graduate Record Examination and the Minority Student

LUIS NIEVES

As the role of standardized tests has increased in the admission process for higher education, more questions have been raised surrounding the effectiveness of these measures for minority and economically deprived students. Because of a lack of relevant data prior to 1974–1975, the Graduate Record Examination Board (GREB), in response to these expressed concerns, cited the findings of studies of other tests similar to the GRE that suggest that, given the traditional nature of a curriculum of academic study, standardized tests predict as well for minorities as for the traditional student (Cleary and Hilton, 1968; Cleary, 1968; Biaggio, 1966). With data on the GRE population now available, the GREB is currently sponsoring research that will more directly test the above hypothesis.

Although the GREB expects that the research will bear out the applicability of the GRE for all subpopulations of those who take the GRE, it wishes to continue to emphasize some limitations of all standardized testing. First, the GREB encourages multiple predictions as more efficient than any single prediction; that is, grade-point average (GPA) with the GRE score is better than the GPA alone. Second, the ability of any instrument or group of predictors to predict depends on certain assumptions about the characteristics of the environment in which a prediction of performance is being made. Specifically, most studies of test validity for the GRE, either alone or in combination with other predictors, have implicitly presumed that the student enters a traditional academic program and that no special intervention, assistance, or support is provided. Once a graduate program adds these

LUIS NIEVES • Program Director, Educational Testing Service, Princeton, New Jersey.

dimensions, the prediction equation is probably weakened, especially for minority students.

In summary, then, the question of the validity of the GRE is dependent on the characteristics of the graduate programs. The GRE anticipates that if graduate programs maintain the same patterns and characteristics, the GRE, in combination with other predictors, is most useful. The decision whether the GRE is applicable to a given (small-core) graduate program may be an equally relevant question. Schools and departments that are interested in using the GRE are encouraged to review sample tests and to evaluate the relevance of both the face and the content validity of the test to their program. If the tests tap the knowledge, abilities, and skills judged to be fundamental to a given graduate program, then it should be used in conjunction with other measures. The GREB is as concerned with eliminating misuse of the GRE as it is in promoting effective, equitable admission procedures and policies.

Institutions using the GRE are advised in the *Guide to the Use of the GRE* to take special care in interpreting GRE scores for minority students. The following is printed in the users' guide:

> Special care is required in interpreting the GRE scores of students who have had an educational and social experience significantly different from that of the traditional majority. These usually include ethnic and racial minorities and students from financially disadvantaged backgrounds. In some cases the differences are compounded by a student's dominance in a language other than English.
>
> Test scores of such students should be considered diagnostic as well as selective and should never be used in isolation. For the most valid estimate of these students' potential, consideration should be given to multiple criteria, some of which may go beyond traditional academic measures. In addition to GRE scores and undergraduate record, evidence of motivation, drive, and commitment to education should be assessed, as well as indications of leadership qualities and interest and achievement in the chosen field of study. (Educational Testing Service, 1975a).

Graduate admissions officers consequently should be both cautious and discerning when evaluating test scores and other information about students who have had educational experiences significantly different from that of the traditional majority of applicants. The general rule that test scores should never be used alone as a basis for admissions or educational decisions should be stringently applied in these cases.

Students are also advised on how graduate institutions use the GRE scores by the following section printed in a brochure sent to every student taking the GRE:

> Although many graduate schools use GRE scores as an aid in assessing the qualifications of applicants for admission or for financial aid, practices vary widely from one school to another and, at some, from one department to another. For example, opinion varies as to the relevance of test scores as measures of the qualities necessary for successful graduate study. Some schools and departments require scores on the Aptitude Test only, some ask just for Advanced Test scores, and others require both.
>
> Practices vary also with respect to the emphasis given scores. The GRE Program staff knows of no graduate school or department that bases its admissions decisions on GRE scores alone. Although some have established minimum test scores and undergraduate averages for acceptability, at many others a high undergraduate average compensates for low scores and vice versa. The most common and best practice is to consider test scores as supplementary to evidence such as the candidate's undergraduate record, letters of recommendation, and other information indicative of fitness for graduate study. In other words, GRE scores are used as only one piece of information in the selection process.
>
> Some graduate schools or individual departments admit all applicants considered qualified for graduate study. However, at many graduate schools, the volume of applications is so great that only some of the best qualified are accepted. Since institutions' requirements, standards, and applicants vary, credentials acceptable at one may not be adequate at others. A verbal ability score of 500 may be satisfactory in one instance and 600 too low for eligibility in a more competitive situation, just as a B average in undergraduate work may be acceptable at one institution and too low for another. Consult your dean, adviser, or the graduate school you wish to attend for further guidance about test scores. (Educational Testing Service, 1975b).

In addition to the Aptitude Test, the GRE program offers a series of twenty Advanced (achievement) Tests in disciplines ranging from biology, chemistry, and education, through Spanish and sociology. These tests are developed by a committee of examiners recruited, with the cooperation of the relevant professional association, from the faculty ranks of institutions of higher education. Most committees are composed of five members, each an expert in some aspect of the field. The committee is charged with designing some of the test questions—others are written by faculty at other institutions—and with selecting questions and setting test specifications. Test specialists at the Educational Testing Service then work with these faculty committees to ensure that the technical requirements concerning reliability, speed, validity, and level of difficulty are met.

The GRE program began collecting self-reported data in October of 1974 and as a result has assembled, for the first time, some specific information about those students who take the GRE. In 1974–1975, 240,836 students (citizens only) took the GRE. Of that number, 7.1% were black; 2.0% American Indian; 1.3% Mexican-American; 1.1% Asian-American; 0.5% Puerto Rican; and 0.5% Latin American (see Table 1). There are some obvious weaknesses in these data that, for the present, limit their confidence level. Because the data are self-reported, they are subject to occasional, but important, discrepancies. For example, according to the data, 4,910 students identified themselves as American Indians. Given past experiences and general enrollment data, it is unlikely that there would be this many Indian students taking the GRE.

A second weakness may be a temporary one. Because the current available data were collected for only one year's GRE administration, we do not know how reliable they are. Finally, we do not know how representative those who take the GRE are of the graduate school applicant population, or of the admittees. The data are nevertheless based on 300,000 test takers, of whom approximately 30,000 report themselves as minority students. This is a large, and in itself an interesting, population without consideration of the question of representation.

Table 2 offers, for selected science fields, the number and the percentage of the minority population reporting the field as their intended major. A blank means that not a single student from the particular ethnic group reported that field as his or her intended graduate major. Chemistry is one of the more popular sciences, with a total of 4,445 candidates reporting that field as their intended major; 189 black students and 75 Asian-American students reported chemistry as their intended major. Combined, these two groups represent 5.9% of all the chemistry students. Combined, all the minority groups represent 7.8% of all the chemistry students. Anatomy, as you can see, is not popular

Table 1. Minority Students Taking the GRE: Citizens Only

	American Indian	Black	Mexican-American	Asian-American	Puerto Rican	Latin American	White	Other	NA[a]	Total
Male	2,774	6,353	1,652	1,407	598	596	102,750	3,197	—	119,327
Female	2,136	10,737	1,432	1,292	597	569	102,909	1,837	—	121,509
Total	4,910	17,090	3,084	2,699	1,195	1,165	205,659	5,034	0	240,836
% of GRE Population	2.0%	7.1%	1.3%	1.1%	0.5%	0.5%	85.4%	2.1%	0.0%	100.0%

[a]NA = No Answer

Table 2. GRE Test Takers by Selected Fields

Fields	American Indian		Black		Mexican-American		Asian-American	
	No.	%[a]	No.	%[a]	No.	%[a]	No.	%[a]
Agriculture	50	2.8	80	4.5	9	0.5	13	0.7
Anatomy	1	3.0	—	—	—	—	—	—
Audiology	5	1.8	21	7.6	1	0.4	5	1.8
Bacteriology	5	2.2	9	4.0	3	1.3	19	8.4
Biochemistry	18	2.1	6	0.7	1	0.1	40	4.6
Biology	247	1.8	661	4.8	109	0.8	214	1.6
Biophysics	4	5.6	1	1.4	—	—	3	4.2
Botany	5	0.9	4	0.7	3	0.5	13	2.2
Chemistry	62	1.4	189	4.2	28	0.6	75	1.7
Entomology	1	0.7	1	0.7	1	0.7	2	1.4
Genetics	4	3.0	—	—	1	0.7	5	3.7
Mathematics	116	1.8	447	6.9	51	0.8	101	1.6
Microbiology	21	2.0	25	2.3	13	1.2	26	2.4
Parasitology	—	—	—	—	1	10.0	—	—
Pathology	1	1.2	8	9.6	—	—	1	1.2
Pharmacology	—	—	1	5.3	—	—	1	5.3
Physics	34	1.5	49	2.1	9	0.4	41	1.8
Physiology	6	2.6	3	1.3	1	0.4	15	6.5
Zoology	52	1.9	53	1.9	6	0.2	46	1.6
Statistics	3	2.8	2	1.9	1	0.9	4	3.7
Other biological sciences	111	2.1	404	7.8	33	0.6	79	1.5
Total of ethnic groups	746		1,964		271		703	
Total population of ethnic groups taking GRE	4,910		17,090		3,064		2,699	
% of minority GRE takers reporting a science as intended major	15.1%		11.5%		8.8%		26.0%	

[a]Percentages of intended majors in a field who are in the given minority groups.

with minority students. Only 3% of all the minority students reported this field as their intended major. The 3% represents one student.

From this table, we can determine that the total number of minority students taking the GRE for 1974–1975 was 30,043 (citizens only), of whom 3,997 reported their intended major in some field of science.

Students considering taking the GRE are advised in the registration booklet in a section entitled "Preparing for the Tests." The following is and Ethnic Group: Citizens Only, 1974–1975

Puerto Rican		Latin American		White		Other		Total	
No.	% [a]	No.	% [a]	No.	% [a]	No.	% [a]	No.	% [a]
6	0.3	—	—	1,583	89.4	29	1.6	**1,770**	0.7
—	—	—	—	31	93.9	1	3.0	**33**	—
—	—	1	0.4	239	86.9	3	1.1	**275**	0.1
—	—	3	1.3	184	81.8	2	0.9	**225**	0.1
2	0.2	3	0.3	760	88.3	31	3.6	**861**	0.4
60	0.4	56	0.4	12,123	88.4	240	1.8	**13,710**	5.7
—	—	1	1.4	61	84.7	2	2.8	**72**	—
4	0.7	—	—	548	93.4	10	1.7	**587**	0.2
18	0.4	18	0.4	3,970	89.1	95	2.1	**4,455**	1.9
1	0.7	1	0.7	132	91.7	5	3.5	**144**	0.1
—	—	1	0.7	119	88.8	4	3.0	**134**	0.1
20	0.3	19	0.3	5,636	86.9	93	1.4	**6,483**	2.7
4	0.4	4	0.4	965	89.8	17	1.6	**1,075**	0.4
—	—	—	—	9	90.0	—	—	**10**	—
1	1.2	—	—	68	81.9	4	4.8	**83**	—
—	—	—	—	15	78.9	2	10.5	**19**	—
5	0.2	9	0.4	2,123	90.7	70	3.0	**2,340**	1.0
—	—	1	0.4	195	84.8	9	3.9	**230**	0.1
17	0.6	8	0.3	2,572	91.6	53	1.9	**2,807**	1.2
2	1.9	2	1.9	94	87.0	—	—	**108**	—
21	0.4	25	0.5	4,395	84.8	117	2.3	**5,185**	2.2
161		156		35,822		787		**40,606**	
1,195		1,165		205,659		5,034		**240,836**	
13.5%		13.0%		17.4%		15.6%		**16.9%**	

the actual text:

Special study for the verbal section of the Aptitude Test is not likely to be effective. If you have not used mathematics for some time, you may wish to prepare for the quantitative questions of the Aptitude Test by reviewing basic algebra and geometry. No advanced mathematics is required for the Aptitude Test.

When preparing for an Advanced Test, a general review of your college courses in the subject may help you organize your knowledge. The descriptive booklet sent with your admission ticket is the most authoritative and up-to-date source on the scope and content of that particular Advanced Test.

> A number of commercially sold publications purport to help candidates improve their scores on the Graduate Record Examinations. Educational Testing Service has reviewed several publications and has found that few of the questions resemble those in the GRE, either in their educational level or in the knowledge required to answer them. Because so many of these publications are misleading, the GRE Program has made available to students a full-length sample copy of the GRE Aptitude Test. You may wish to purchase a copy to familiarize yourself with the type of questions asked. Information on how to order the sample Aptitude Test is on page 10. Full-length samples of the Advanced Tests are not available. (Educational Testing Service, 1975b)

Investigators have conducted research to determine if it was possible to affect (increase) performance on standardized tests by short-term coaching or instruction (Roberts and Oppenheim, 1966). Generally, the results have not shown that coaching and/or instruction will pay off in increased performance. Criticism of these studies has suggested that the populations used were either too sophisticated or too academically deficient (Evans and Pike, 1973). Studies are continuing to be sponsored to ensure that the test is not subject to score fluctuation from short-term cramming, coaching, or instruction.

Since the GRE Aptitude Test measures skills developed over a long period of time, it is not surprising that almost all studies conducted have so far verified the unsusceptibility of the verbal GRE to coaching. The quantitative portion utilizes only basic geometry and algebra principles. If these are not known, they can, of course, be taught. We suspect that if any significant variation is possible on the quantitative GRE, it will be on only the low-performance students and it will be slight. Generally, the GRE is maintained in design, form, and content so that effects of short-term coaching/instruction are eliminated or reduced to negligible differences. In summary, we believe that on the whole, neither the verbal nor the quantitative test of the GRE is particularly subject to increments due to coaching—the quantitative, because of the limited number of math principles tapped. There may be item variations, but they probably do not affect the overall score. There is, however, some possible value in test *familiarity*—as opposed to trying to *learn* the skills involved in the Aptitude Test—and for this reason, the GREB has made available full-length sample copies of the Aptitude Test. And, of course, scores on the Advanced (discipline) Tests can be improved by increased knowledge in a discipline; the extent to which such an improvement can be accomplished in a short-term cram course is problematical at best.

Although the GRE verbal and quantitative aptitude scores have proved to be reliable and important aids in the admission process, the GREB is now investigating ways to make the Aptitude Test more useful to students and institutions, primarily by including measures of characteristics as yet untapped by the GRE. The purpose of contemplated change is fourfold: (1) to broaden the definition of scholastic talent, thus enabling students with abilities relatively independent of verbal and quantitative skills to demonstrate their strengths; (2) to tailor portions of the test to groups for whom the measures are most appropriate, thus individualizing the test as far as possible in a standardized setting; (3) to increase the predictive validity of the test; an (4) to provide for the possibility of future guidance instruments to help applicants make better choices of programs and careers.

Although change is being proposed, the board is assuming that constant measures are also desirable. Thus, according to current plans, verbal and quantitative scores on the current scale and with the current meaning will continue to be reported for all candidates. If research proves that modifications can enhance the test's usefulness, and if users welcome the modifications, the Aptitude Test will undergo three basic changes in the near future: (1) the verbal and quantitative measures will be shortened to allow time for an additional measure; (2) the broad-ranging reading-comprehension section will be replaced by two reading options (one on scientific topics, the other on humanities and social-science topics); (3) a measure of abstract-reasoning skills will be added to the Aptitude Test.

Current research is related to the proposed test model and will focus on such questions as the reliability, the validity, and the relative independence of proposed new item types measuring abstract-reasoning skills. Investigators will also obtain information concerning the comparability of currently reported scores and their counterparts in the new model.

Proposals to modify the Aptitude Test assume that it should be an instrument responsive to changing needs and test capabilities. Thus, a measure of scientific thinking (covering such skills as formulating hypotheses, drawing inferences from data, and evaluating results of experiments) and an index of "cognitive style" or approach to study that may eventually lead to a guidance instrument related to choice of career and academic discipline are now in the early stages of research.

REFERENCES

Biaggio, A. Relative predictability of freshmen grade point averages from SAT scores in Negro and white southern colleges. *Technical Report #7,* Research and Development Center for Learning and Re-education, University of Wisconsin, Madison, 1966.

Cleary, A. T. Test bias: Prediction of grades of Negro and white students in integrated colleges. *Journal of Educational and Psychological Measurement,* 1968, *5,* 115–124.

Cleary, A. T., and Hilton, T. L. An investigation of item bias. *Education and Psychological Measurement,* 1968, *28,* 61–75.

Educational Testing Service. Guide to the use of the Graduate Record Examinations, 1975–77. Princeton, N.J.: ETS, 1975a.

Educational Testing Service. What your GRE scores mean, 1975–77. Princeton, N.J.: ETS, 1975b.

Evans, F. R., and Pike, L. W. The effects of instruction for three mathematics item formats. *Journal of Educative Measurement,* 1973, *10,* 4.

Roberts, S. O., and Oppenheim, D. B. The effect of special instructions upon test performance of high school students in Tennessee. ETS RB-66-36. Princeton, N.J.: ETS, 1966.

II. Problems of Minorities at Majority Institutions

5

Minority Students and the Political Environment: A Historical Perspective

THEODORE M. BROWN

I feel like a strange interloper in these proceedings. Most other conference participants are either educators with long practical experience in training minority students in the sciences and medicine or social scientists who have studied the attendant problems for years. By contrast, I am an historian, an historian of science and medicine more specifically, and I am very new to the business of medical education. Although I have been the assistant director of the Center for Biomedical Education (CBE) at the City College of the City University of New York (CUNY) for a little over two and a half years and in that capacity have performed many functions, I serve primarily to represent the social sciences and humanities in the center's unusual educational curriculum and to embody what might be called a societal or ethical perspective.

Having said this much to disarm you, let me now double back and suggest why my presence here is not entirely accidental or, I hope, inappropriate. During my two and a half years at the CBE, we have lived through an extraordinarily turbulent history. We have been subjected to enormous political pressures: from black, Jewish, Puerto Rican, and Italian civic and community groups and from every stripe of political figure. We have had to live with the suspicion and often the hostility of faculty colleagues, some administrators, and other students on a vast, troubled urban campus located in Harlem and cherished by frequently influential and powerful alumni who remember the City College of a different era and style. Nathan Glazer (1975), the Harvard sociologist who has just published a new book seriously questioning the ethics and efficacy of affirmative-action programs, is a City College

THEODORE M. BROWN • Assistant Director, Center for Biomedical Education, City College of the City University of New York, New York.

alumnus. So is, to suggest a different range of concerns, Mayor Abraham Beame.

The scrutiny to which we have been subjected has not been dispassionate nor from afar. Some of our critics have confronted us directly, and others have received much of their information through distorted and sensationalized accounts in the metropolitan press. When, for example, we allowed certain students to take a reexamination in one of our courses, this "scandal" was reported with an inch-and-a-half headline on page three of *The New York Post*. For a while it seemed as if the *Daily News* were running a weekly feature story, also on page three, on the latest "scoop" from the CBE. *The New York Times* did its share too. Most recently, we have been hit with class-action lawsuits by the Anti-Defamation League and the Italian American Center for Urban Affairs. After preliminary legal maneuvers of many months, trial is about to begin in the Federal District Court of our alleged "reverse-discrimination" case. All this while we have been trying to build a novel form of premedical and medical education with a limited budget and a far-from-excessive staff.

We in the CBE have been living an intense history. Perhaps it is time now to record it and to draw its lessons. This seems a good occasion to begin that process, for the history we have experienced and the lessons we have learned will be of interest, at least by analogy, to those concerned with the problems of educating minority students in majority institutions and, more generally, to all those interested in developing nonacademic predictors of success in science and medicine.

The Center for Biomedical Education was formally created in November 1972 and enrolled its first class of students in September 1973. The real starting point of its history, however, was April 22, 1969, when the modern history of City College, of which the history of the CBE is an inseparable part, began. On that day, a group of black and Puerto Rican students took control of City College's "south campus." These students pressed "five demands," among which was the immediate enrollment at the City University of a sufficient number of black and Puerto Rican students to bring their overall enrollment in CUNY up to the percentage level of blacks and Puerto Ricans in the city at large. This demand was widely supported by blacks and Puerto Ricans in the city because it was felt that the tuition-free, publicly supported City University had functioned too long as an elite, essentially white institution despite the massive demographic changes in New York. The demand for what was seen as a "quota" by white ethnic groups and their political spokesmen was bitterly resisted, and a stalemate occasionally interrupted by violence ensued. A compromise, "open admissions," ended the stalemate and established an uneasy peace.*

It came as a surprise to many that those who seemed to benefit most from open admissions were lower-income white Catholics, but blacks and Puerto Ricans certainly entered CUNY in far larger numbers than ever before. In 1969, blacks made up 14% of CUNY's freshman class and Puerto Ricans, 6%; by 1972, the percentages were 27% and 14%, respectively. Partly as a result of long-standing demographic shifts and partly in fear of the sort of violence that was widely reported in 1969, those highly qualified white students who still entered the City University under open admissions increasingly opted for senior-college campuses other than City College. Queens College and Brooklyn College rapidly became enclaves of student "excellence," whereas City College faculty were increasingly faced by very large numbers of poorly prepared students—white, black, and Puerto Rican. In the fall of 1972, large sectors of the City College faculty were at the point of almost complete demoralization. They hoped for some dramatic event that would shunt away the hordes of open-admissions students and "bring the good students back."

Dr. Robert E. Marshak, since September 1970, the eighth president of City College, took decisive action. He announced a series of new programs that would be created to attract talented students back to City College. The first and the major one of these was to be the Center for Biomedical Education, which would offer the first four years of an unusual six-year B.S.–M.D. program. Students would enroll in the center immediately after high school and in the space of four years master a special curriculum encompassing all the basic biomedical sciences normally taught in the first two years of medical school and as many of the standard premedical subjects as were necessary. Thus, in their freshman year, students would have a course on medical-school level in human gross anatomy—without any prior college-level biology—and an intense course in general and organic chemistry. In their sophomore year, students would be ready for biochemistry and would continue the study of human biology with histology. So it would go through the four years of the curriculum. When students qualified for

*For the open-admissions story and some of the information in the next paragraph, see Mayer (1973) and Harrington (1975).

their B.S. degrees in biomedical science and passed Part I of the National Board examinations, they would transfer to the third year at one of a variety of cooperating medical schools to complete their clinical training for the M.D. degree.

The main purpose of the biomedical program, Marshak announced, was not merely to experiment with premedical and medical education by recombining and accelerating it but to achieve broad social goals. Urban areas, especially inner cities, are underserved by primary-care practitioners, and thus the social objective of the CBE was to prepare its graduates for careers as primary-care physicians in underserved urban communities. To help meet this objective, a special sequence of courses and field experiences collectively labeled "Health, Medicine, and Society" was inserted into the curriculum. The other means for achieving the broad social objective of the CBE was to institute an unusual admissions procedure. If the center could somehow find those students who were truly motivated to become primary-care practitioners in underserved urban neighborhoods, then, it was thought, the problem of delivering on its promise would largely solve itself.

Admissions policy and procedures quickly became the storm center of the CBE. In late 1972 and early 1973, admissions was the prized object in a tug-of-war between the various forces at City College, which, in tense equilibrium, kept the campus at the edge of chaos. One tug, of course, came from members of the science faculty, who desperately wanted "the good students back." The CBE would be the magnet that would draw those talented science students to City College once again. Another, very different tug came from minority faculty and staff, of whom there were a goodly number at City College, especially now after 1969. The minority voice was, of course, raised against the reinstitution of "elitist" admissions practices at City College. If the CBE were allowed to admit students on the basis some of the science faculty urged, then the struggle for open admissions would have been for nought. It was feared that City College would soon turn the clock all the way back and undo open admissions. Worse still, the college would be manipulating and "ripping off" communities in desperate need of medical help, because the need of those communities seemed to legitimate the CBE and thus aided in its creation.

Several times during the spring of 1973, it looked as if the CBE would never really come to birth, that it would abort amid the agitation over admissions. But something of a compromise was achieved when it was decided that the admissions committee would pay careful attention to both "academic" and "nonacademic" criteria in selecting the first class of biomedical students. No specific definition was given for these terms and no proportion was suggested for the best balance of academic and nonacademic factors in the assessment of candidates for admission. But politically the compromise worked. Both major forces decided that they could live with the arrangement, probably because each assumed that a balance could be achieved that would be most advantageous to the particular applicants it was interested in. The practical consequence of this nondecision was the agreement by the admissions committee to interview in teams of two *every* candidate for the September 1973 entering class and to grant essentially final admissions authority to each interviewing team.

The results the first semester in the program were what might have been expected, given the initial admissions procedure. A large proportion of the class did quite well and experienced few academic difficulties. Another large portion of the class, however, had major problems. Many began to drop back and several wanted to drop out. They were counseled to stay in. We wanted and expected everyone to get through. Reexaminations were scheduled in certain cases after students had time to master material they missed the first time around. Faculty were nervous and the CBE administrators were confused, because City College students are not usually allowed a "second chance" in their courses. Were not we favoring CBE students with special privileges, whereas other, perhaps more academically talented City College students had to sink or swim in their courses, while their chances at medical-school admission depended on their one-shot success. The upshot of this turmoil was that the decision to reexamine certain students—a procedure that was already commonplace in medical schools—became something of a "scandal" at a highly politicized, ethnically conscious, competitive City College. Word of our academic procedure was leaked to the student press and from there found its way into *The New York Post.* We in the CBE felt foolish. Instead of unflinchingly defending our students and the propriety of our procedures, we began to talk about the "conformity" of our academic policies with those of the college at large. We also began to consider criteria for the academic "probation" and "separation" of our poorer students from the CBE. We held meetings with the provost and other college officials to learn in more detail about customary grading, testing, and probationary actions. This was not public relations; we had become seriously concerned.

Meanwhile, a storm was brewing over admissions. Several disappointed applicants for the September 1973 class had lodged complaints—with us, with the Anti-Defamation League, and with political figures. We responded, explained, cajoled. No one was really satisfied. We were threatened with legal action, and throughout 1973 and 1974, many efforts were made to head off formal legal proceedings. But the Board of Higher Education, also under pressure, gave the CBE a deadline by which to reform its admissions process. Nevertheless, after months of discussion and delay, the Anti-Defamation League still went to court, joined by the Italian American Center for Urban Affairs. We are just now about to come to trial.

The ADL's impending lawsuit has been the principal element in the atmosphere in which the CBE has had to function for the last year and a half. We have been constantly aware of external political-cum-legal pressures. We have been aware, too, that external political fluctuations have had resonances in the internal campus environment. Thus, those members of the faculty who never understood or agreed with our early admissions criteria have had their resolve strengthened by the ADL legal action and the surrounding publicity. These days, minority faculty and staff make few public pronouncements. The campus political equilibrium has shifted markedly.

The effects of the new political pressures to which we are subjected are not difficult to discover. Without always realizing it at the time, we have responded in what may seem in retrospect a defensive manner. With regard to admission, we have moved away from the attempt in 1973 to assess the motivation, community orientation, and maturity of every applicant. In 1974, guided largely by pragmatic considerations, we decided to interview only the academically best-qualified half of the applicant pool. But in 1975, we began to respond to political as well as practical necessities: much greater weight overall was placed on academic qualifications than in the previous year. In 1975, we interviewed only 300 candidates. In 1976, we will interview 250.

As the number interviewed decreases, the influence of academic criteria increases. We have, in other words, gotten more into the position of favoring for admission those students who have the best chance of surviving our curriculum academically. Although we always had an academic cut-point and we still consider nonacademic factors seriously, those with superb nonacademic qualities have a harder time getting admitted now than they did before. And as we alter our admissions process, we put progressively more faith in our Health, Medicine and Society curriculum as a source of motivation. This is doubtlessly appropriate, but it certainly represents a change in attitude. These shifts are subtle, and we don't often pause to reflect on them or their implications.

These same basic patterns are evident too in our academic progress and retention procedures. In the early days of the CBE, we genuinely hoped and perhaps even naïvely expected that everyone admitted would get through. Great drive and motivation, with proper supports, could be imagined to compensate for whatever deficiences there may have been in academic preparation. Today, however, our rules have become rather strict. Our expectations have changed. We can no longer easily imagine drive and motivation compensating for every sort of academic deficiency. We have begun to act on logic that increasingly puts us in the position of selecting students who can manage the curriculum academically rather than designing an admissions procedure that selects for and a curriculum that enhances the best qualities students bring in with them.

The story I have told about the CBE is a story about politics and perception. I have tried to suggest how views on what is real and what is important shift as pressures shift and the balance of forces displaces from one equilibrium position to another. There is, however, another interpretation of the events I have recounted. A different observer might have concluded after reviewing the brief historical record that we simply passed from ignorance and naïve enthusiasm to knowledge and mature judgment. It might be said that we in the CBE have merely learned through experience that motivation is difficult and perhaps foolish to assess accurately; that only students above certain academic thresholds are capable of performing well in premedical or medical-school courses, whatever their motivation; that long-range academic success or failure can be determined early in a student's career; and that it is better either not to admit or to "counsel out" the likely failures than to prolong their unfounded dreams. There certainly is that rational thread through our history, and it is probably the most substantial one. But I have learned in my studies as a historian of science and medicine that interpretations of rationality are usually more complicated and that rationality and environmental pressures are rarely, if ever, completely separate. Historical instance after instance shows that scientists and physicians have adopted new criteria of rationality, new ways of interpreting old data, when intellectual and political environments changed. As just one example of many striking ones that might be cited, it now seems well documented that Isaac Newton formulated or at least clarified some of his basic ideas about force in response to the political and theological turmoil of the English Revolution and in the desire to distinguish his view of the universe from that of heretical materialists (Rattansi, 1973). He was probably unconscious of his full range of motives and would have responded with rationality if asked for his justification. Clearly, the greatest scientific minds, even when they are not aware of it, do not function completely free of the influence of environmental circumstances.

What, then, are the implications of all this for an understanding of the problems of minority students at majority institutions and the particular influence of the political environment? How does a review of the CBE experience and the drawing of general historical lessons help us understand the present and formulate policies for the future? Let me use my remaining time to suggest ways in which a historical perspective might be helpful to an understanding of the situation of minority students (primarily blacks) at majority medical schools, that is, all those in the United States except Howard and Meharry.

I must first sketch in very brief outline the trends of minority-student admission to American medical schools. Most of my data are taken from Dr. James Curtis's *Blacks, Medical Schools, and Society* (1971) and recent papers published in the *Journal of Medical Education* (Davis and Sedlacek, 1975; Davis, Smith, and Tarnoff, 1975). I will also cite as yet unpublished data generously supplied to me by James Angel, Director of the Medical College Admissions Assessment Program at the Association of American Medical Colleges (AAMC).

Briefly, the basic trends are these. In 1938–1939, there were 350 black students enrolled in all United States medical schools, including Howard and Meharry. This represented 1.64% of the total medical-school enrollment. Only 12.9% of the 350 black students that year were enrolled in predominantly white schools. By 1950–1951, there were 651 black students in all American medical schools. This represented 2.52% of the total, an increase probably due in part to pressure from the NAACP and a changing legal and political environment. Nonetheless, in 1950–1951, only 21.6% of the black students were enrolled at predominantly white institutions. By 1968–1969, the percentage of black medical students in the expanding overall medical-student population had dropped to 2.18%, although 37.3% of these were in predominantly white institutions.

Given the political circumstances of the late 1960s, many felt that something major had to be done. In 1968, the recently formed AAMC Assembly had adopted a recommendation declaring, "Medical schools must admit increased numbers of students from geographic areas, economic backgrounds and ethnic groups that are now inadequately represented" (Davis *et al.*, 1975, p. 721). In 1969, the AAMC received a grant from the U.S. Office of Economic Opportunity to "increase educational opportunities for minorities in the health professions," and in 1969–1970, an AAMC task force looked into the representation of minority students in medical schools. The task force report was published on April 22, 1970 (coincidentally, exactly one year after the City College south-campus takeover by black and Puerto Rican students) and recommended that United States medical schools aim for at least 12% black freshman enrollments by 1975–1976. This task-force report was endorsed by the AAMC, the American Hospital Association, the American Medical Association, and the National Medical Association.

Beginning in 1969–1970, the picture changed substantially. The first result of the new political equilibrium in that year was the marked increase of minority-student enrollments. Already in that year, the number of black medical students jumped from the 2.18% of the previous year to 2.75%, 52.4% of whom were now enrolled in predominantly white schools. The bulk of this increase, obviously, had to come as the result of new freshman admissions. By 1971–1972, black freshman enrollments represented 7.1% of new medical enrollments. All "underrepresented" minorities (including also American Indians, Mexican-Americans, and mainland Puerto Ricans) represented 8.6% of the first-year enrollment totals. By 1974–1975, minority representation in the freshman class had increased to 10.0%, 7.5% of this owing to black students.

From one perspective, the picture was suddenly looking very good. But there are other perspectives too. It has been pointed out that after the initial spurt, minority-student enrollments quickly leveled off. Black students in particular have never reached beyond 6.3% of total medical-school enrollment. Indeed, it is perhaps most frightening to realize that just-prepared and as-yet-unpublished data for 1975–1976 indicate a *drop* in black freshman enrollment from 7.5% the previous year to 6.8% this year. Both figures are well below the 12% goal for 1976 set in 1970.

What accounts for this apparent pattern, assuming that it is not too early to tell about new or developing trends? There are obviously many factors, which include the number and characteristics of minority premedical applicants, but these factors are too complicated for me to analyze here. Suffice it to say that a clearly changing political environment, which may well be affecting medical schools and medical-school admissions committees, may be one very large factor in the picture as of 1976.

What are the elements of the new national atmosphere that are likely affecting minority admissions to medical school? In a sense, these are probably not very different from the elements in the atmosphere at City College. First, we have, of course, passed from a period of civil rights and affirmative action to a period of white ethnicity and strongly expressed doubts about affirmative programs in employment and admissions. Second, many in the medical-school community are getting nervous, perhaps excessively so, over actual or threatened legal action contesting admissions procedures and criteria.* The DeFunis case is clearly part of this environment, and other cases are obviously in the offing. Finally, of course, there are the facts of minority-student academic profiles upon admission to medical schools with undeniably lower grade-point averages and MCAT scores than vigorously competing white applicants. Moreover, it seems likely that minority students have more difficulty once in medical school than do nonminority students. Whereas 97% of the white students who entered in 1971 remained in medical school, the figure for black students seems to be 91% and may even be lower (Davis *et al.*, 1975, p. 738).

What are medical-school faculties to do with such information? Should they throw up their hands and admit failure in the experiment of affirmative action with regard to the enrollment and retention of minority students, or should they acknowledge instead that the problem of minority-student preparedness for medical school goes well back beyond the threshold of entry? College, high-school, and perhaps even primary-school educational environments clearly contribute to good or

*For an important response to the "excessive" nervousness of medical school admissions officers, see Begun (1973).

poor performance on standardized tests and in medical-school curricula. What can be done to improve these environments to make them more conducive to minority-student success in medical school?

Whatever we attempt, it will be valuable to maintain a clear historical perspective. When we monitor our attitudes, vocabulary, and overt behaviors, can we clearly detect when we are operating on the basis of well-informed, rational judgments as opposed to subtle political motives that may be hidden from our own scrutiny? When we say that Student A, who is experiencing difficulty in the curriculum, has been "admitted with insufficient academic preparation," do we really mean this, or do we mean that, given our budgetary and other priorities, we haven't decided to provide him or her with sufficient assistance when he or she arrived at medical school? When we worry about the ethics of denying admission to apparently better-qualified majority students in order to make room for minority students, are we truly worried about ethics or about threatened litigation and adverse publicity? Do we keep clearly in mind that we are looking for a variety of qualities in our applicants and not merely high averages and test scores? Are the social goals and academic sensitivities clearly perceived in 1970 being lost sight of as the decade, with its changing politics, proceeds?

These questions are surely difficult to answer and perhaps even unfair. But imagine this situation. In 1980, American society with its political and economic realities suddenly changes so dramatically that medicine is no longer an attractive career for young Caucasian men and women. They start going by droves, let us say, into the suddenly socially prestigious and lucrative profession of cultural anthropology. As a result, the number of majority-student applications to medical school dwindles to near zero. Political spokesmen for the majority students begin to criticize graduate schools for restricting the admission of students to cultural-anthropology training programs. Lawsuits are threatened against "preferential" admissions practices in several departments that seem to favor American Indian applicants. Admissions officers panic and begin to consider admitting students only on the basis of GRE scores. Meanwhile, minority students still continue to apply in reasonable numbers for medical-school admission. Is there any question that larger numbers of them would win admission in hypothetical 1980 and, once admitted, would progress toward their medical degrees? Some of the newly admitted medical students might find the going rough. Is there any question but that soon after 1980, medical-school faculties would adjust their pace of instruction, offer special programs in plethora, and perhaps even think about the unthinkable—altering the curriculum?

My example is absurd, but it makes a point. Can we ever really separate political considerations, which in turn depend upon social and economic factors, from other considerations used in the assessment of students for admission or retention in highly competitive programs? Since, I think, we cannot, let us at least be aware of the presence and influence of politics wherever we find it, and let us be prepared to exhibit it to all who will see.

REFERENCES

Begun, M. S. Legal considerations related to minority group recruitment and admissions. *Journal of Medical Education,* 1973, *48,* 556–559.

Curtis, J. *Blacks, medical schools, and society.* Ann Arbor: University of Michigan Press, 1971.

Davis, G. J., and Sedlacek, W. E. Retention by sex and race of 1968–1972 U.S. medical school entrants. *Journal of Medical Education,* 1975, *50,* 925–933.

Davis, G. J., Smith, V. C., and Tarnoff, S. L. Recruitment and progress of minority medical school entrants 1970–1972. *Journal of Medical Education,* 1975, *50,* 713–755.

Glazer, N. *Ethnic inequality and public policy.* New York: Basic Books, 1975.

Harrington, M. Keep open admissions open. *New York Times Sunday Magazine,* November 2, 1975, 16ff.

Mayer, M. Higher education for all? *Commentary,* 1973, *55* (2), 37–47.

Rattansi, P. M. Some evaluations of reason in sixteenth and seventeenth century philosophy. In M. Teich and R. Young (Eds.), *Changing perspectives in the history of science.* Boston: Reidel Press, 1973. Pp. 148–166 (esp. *re* Newton, pp. 154–165).

6

Status and Prospects of Minority Students in Traditional Medical Schools

LEROY T. BROWN

There is a need to increase the number and the success rate of minority students enrolling in medical careers. Although an increasing number of minorities enter the freshman year of undergraduate education in Wisconsin, the rate at which they enter is still substantially less than that for all students in the state as a whole (USDHEW, 1970, 1972a,b; see Table 1).

The likelihood of minority students' continuing to graduate school is also less than that for nonminorities, and this is particularly true for medical school. Nationally, in 1970, there were roughly one-half million students enrolled in graduate school, representing 10.3% of the undergraduate enrollment (see Table 2). The rate at which blacks were entering the medical schools was even less impressive, although since 1970, there has been a steady increase in the numbers of minorities enrolled in medical school: from 1,723 in 1970 to 4,323 in 1974. However, the National Medical Fellowships (1975) reported that 1975–1976 showed the first drop in the size of first-year minority enrollment: from approximately 1,600 in 1975 to 1,400 in 1976. In some part, this drop has been due to the sharp cutback in the federal funding of medical education. The National Medical Fellowships reported that it is unrealistic to believe that medical schools will continue to recruit minority students as vigorously in view of decreasing financial resources.

In some states, such as Wisconsin, early recruitment of minority students was not aggressive, as evidenced by the fact that since 1969, the University of Wisconsin has graduated only 15 minority medical students. In 1970, there was a total of 5 minority medical students

LEROY T. BROWN • Assistant Vice Chancellor for Health Sciences, Health Science Opportunity Programs, University of Wisconsin, Madison, Wisconsin.

Table 1. Racial and Ethnic Enrollment, Fall 1970 and 1972[a]

State of Wisconsin	Elementary and secondary education	Undergraduates	Percentage
Total			
1970	987,917	137,426	13.9
1972	984,326	136,530	13.8
Minority			
1970	59,295	4,433	7.4
1972	63,914	5,421	8.5

[a]Data compiled from U.S. Department of Health, Education, and Welfare, Office for Civil Rights (1970, 1972a).

enrolled in the four years of the University of Wisconsin Medical School, representing 2.5% of a total enrollment. The national enrollment of minorities that year was 4.3%. In 1972, the faculty went on record as favoring the establishment of a substantive minority-student program. Today, there are 36 minority students enrolled in the University of Wisconsin Medical School, representing 6% of the total enrollment.

The slow pace of minority enrollment in medical school in Wisconsin is in large part due to the sparse potential-applicant pool. In a total population of over 4 million, there are only 5 counties in Wisconsin that have black populations greater than 1,000, and of the remaining 64 counties, only 9 have black populations between 100 and 500 (U.S. Bureau of the Census, 1970). There are many counties in Wisconsin that have relatively few minority members, and some that have none.

In 1970, the Association of American Medical Colleges (AAMC) Task Force on Expanding Educational Opportunities in Medicine for Blacks and Other Minority Students (AAMC, 1970) projected that 6% (5,100) of the 85,000 minority students that had entered the freshman year of college in 1968–1969 would express an interest in medicine as a career, and that in 1972, 25%, or 1,275 of the 5,100, would actually

Table 2. Undergraduate–Graduate Enrollment, 1970[a]

National	Undergraduate	Graduate	Percentage	Medical school	Percentage
Total	4,965,768	512,004	10.3	40,238	0.8%
Black	464,734	29,926	7.1	2,582	0.5%

[a]Data compiled from AAMC (1970, 1972).

apply. They felt that the chances for a minority applicant's admission was 75% as compared to a 45% probability for nonminority applicants. What actually happened in 1972 was that blacks alone constituted twice the projected number of minority members who applied to medical school. Medical schools, however, became much more critical in their admissions criteria and accepted only 36%, or 857 blacks.

Using the task-force projections to estimate the potential minority-applicant pool in Wisconsin, we can anticipate that of roughly 6,000 minority undergraduate students in all four years, approximately 120 minority students might express an interest in medical school, and approximately 30 will actually apply. The two medical schools in the state could expect to enroll 15 or 16 minority students each. Although the state, as a whole, has a minority population of 4.5% (1970 census), there are relatively few minority physicians. (The largest concentration is the Cream City Society of minority physicians in Milwaukee.) An entering class of 15 students is a modest number.

If the admissions criteria are not to be lowered, considerable effort must be made in developing the potential minority-applicant pool for any significant increase in the numbers of minority students admitted to medical school in Wisconsin to occur. Without question, greater interest in medical careers must be developed in minority youth during the early years of high school and possibly earlier. The sophistication of modern-day medicine demands a strong preparation in hard sciences, that is, mathematics, physics, and chemistry, as well as molecular and cellular biology. At one time, 20 years ago, comparative anatomy was the advanced biology course that determined which premed majors would enter medical school. Today, the course in cellular biology—that is, the metabolism of organ systems at the subcellular level—is involved in that decision. It is essential that students develop good mathematical and scientific skills, as well as reading skills, early in their high-school years. The course that seems to present the strongest challenge to first-year medical students is physiological chemistry. It is here that the students' preparation in advanced mathematics, quantitative analysis in chemistry, and advanced cellular biology (i.e., a chemistry course in disguise) comes to greatest use. A high-school student excelling in biology but not in mathematics and chemistry very rapidly experiences deficiencies in medical school.

Of those enrolled in the University of Wisconsin Medical School, 95% are from the state of Wisconsin, as mandated by the legislature. Of these, 40% are students enrolled in or graduates of the Madison campus. Of accepted medical students, 60% are biology majors. The Madison campus has developed an advanced biology curriculum called the biocore track, which includes some premedical courses. The track is designed to challenge high-achieving students. A strong command of advanced mathematics and calculus is included in the prerequisites. Few minority students are enrolled in these courses.

All this is not to underestimate the need for students to have a good knowledge and interest in biological systems. It must be realized, however, that the academic record in the premedical curriculum is an important consideration of the candidate by the admissions committee. In Wisconsin, we receive approximately 1,600 applications each year for medical school; several hundred of these are from minority students. Across the nation last year, roughly more than 45,000 students applied for 15,000 places. It is estimated that more than two-thirds of those who applied will not gain entrance into medical school this year. More disturbing is the possibility that one-half of those whom we reject are academically qualified. The AAMC realizes that the competition is keen and has begun to develop a test (1975)* that will hopefully identify the qualities, in addition to the academic one, that make a good practicing physician. Such qualities as compassion, sensitivity, and awareness (i.e., a general sense of appreciation of people and health-care delivery) become important. Such noncognitive qualities may be difficult to test or to present on a quantitative basis. Perhaps such an examination could explore conflicting value situations. For example, low-income people, poor people, minorities in general, and blacks in particular do not have proper access to good medical care. They often go from one emergency room to another in search of medical care and frequently see only an intern. A recent study (Cheng, 1974) in Illinois revealed that there are fewer black physicians practicing in Chicago today than there were five years ago. In spite of this lack of adequate medical care, a number of the best health-care-delivery institutions in the nation are located in the heart of high-density, low-income areas!

An immediate goal for a health-careers special-project program would be to orient or to direct minority students by way of intensive academic support toward the advanced courses. The program should be carefully structured so as to avoid the possibility of developing a

*The Medical College Admissions Assessment Program (MCAAP) *Cognitive* section will be administered for the first time in the spring of 1977. (See James Angel, this volume.)

pattern of repeated academic failure in the student who is poorly prepared in both study skills and basic knowledge.

Such a program is now under way at the University of Wisconsin. The Office of Educational Resources in the Center for Health Sciences, in cooperation with the basic science faculty, is providing academic support in the form of the study skills and the basic information essential for success in the first two years of medical school. There is a strong indication that academic support of this kind is extremely helpful, particularly in the basic sciences.

During the past two years, minority-recruitment efforts at the University of Wisconsin at Madison have been intensified. There has been a significant improvement in grade-point average and MCAT performance in the quantitative and science areas among entering classes of minority students. Although the attrition at the University of Wisconsin Medical School at Madison has been relatively low, the number of minority students who have had to take an additional year has been proportionately high. It is encouraging, however, to note that the number of minority students making regular and uninterrupted progress through medical school has increased. The increased retention rate is due, in large part, to the greater sensitivity on the part of the teaching faculty to the needs of minority students.

REFERENCES

Association of American Medical Colleges. Task Force Reprint to the Inter-Association Committee on Expanding Educational Opportunities in Medicine for Blacks and Other Minority Students, April 22, 1970.

Association of American Medical Colleges, Division of Educational Measurement and Research. Introduction to Medical College Admissions Assessment Program 1977 Admissions Tests. Washington, D. C.: MCAAP Information Series No. 1, March 1975.

Association of American Medical Colleges, Division of Student Studies. AAMC Enrollment Data, 1970–1975.

Cheng, R. Black physician survey report. Comprehensive Research and Development, Inc., 1974.

Johnson, D. G., Smith, V. C., Jr., and Tarnoff, S. L. Recruitment and progress of minority medical school entrants 1970–1972: A cooperative study by the SNMA and the AAMC. *Journal of Medical Education*, 1975, *50* (Supplement), 713–755.

National Medical Fellowships, Inc. Report of the Scholarship Program. New York: NMF, November 7, 1975.

U.S. Bureau of the Census, 1970.

Department of Health, Education, and Welfare, Office for Civil Rights. Racial and ethnic enrollment data from institutions of higher education, OCR-72-8. Washington, D.C., Fall 1970.

Department of Health, Education, and Welfare, Office for Civil Rights. Racial and ethnic enrollment data from institutions of higher education, OCR-74-13. Washington, D.C., Fall 1972a.

Department of Health, Education, and Welfare, Office for Civil Rights, Public Information Office. Racial and ethnic enrollment in public elementary and secondary schools. Washington, D. C., Fall 1972b.

7

Sociocultural Factors as a Deterrent to Students' Pursuit of the Sciences

SIGFREDO MAESTAS

Much of the area of the southwestern United States is one of vast expanses, clear blue skies, and impregnable solitude. Young women and men in some localities of this region can live their formative years only moderately influenced by the Atomic Age. In many instances, fathers and mothers of the current generation of college students in this area came to know the Industrial Revolution firsthand only during and immediately following World War II. Among these inhabitants of the Southwest are Mexican-Americans and American Indians, who formerly constituted a majority of the population of this largely uninhabited land.

A large number of factors in the current day impinge on the abilities of the young people who are attempting to acquire education in our colleges and universities. There exists a sufficient amount of information in print on the general subject for anyone who wishes to familiarize himself or herself with the educational problems of minority students. One subject that often is not treated is the thesis of this paper: an exploration of social and cultural factors that serve to prevent more Mexican-American and American Indian students from seeking a life's work in science.

The approach to this subject is taken from the author's experience with the myriad of problems that *some* minority students encounter. It is necessary to say that while some phenomena and observable principles alluded to in this discussion may easily apply to all minority students, the author's experience is primarily with rural Mexican-

SIGFREDO MAESTAS • Dean of Academic Affairs, New Mexico Highlands University, Las Vegas, New Mexico.

Americans of the Southwest and with American Indians of the same region.

The social and cultural characteristics of these native groups of the Southwest often become an impediment to the pursuit of education and training in science. The barriers to science for Mexican-Americans and American Indians are numerous. Educators and scientists in this society are familiar with difficulties that students encounter if English is the students' second language. (The tendency is generally to view the problems as cognitive or dictional.) Similarly, teachers are familiar with students' lack of mathematical skills. Well-meaning teachers and educators find it logical to attribute problems that students have with the English language and mathematics to inadequate prior schooling, disadvantaged homes, and, occasionally, lack of motivation. It should follow, then, that our ability to remedy problems of "nonrigorous" education, poor homes, and lack of student interest would provide the prescription to produce more scientists from among young women and men with these backgrounds. Unfortunately, the problem is more complex than that.

The objective here is to illustrate a set of dilemmas that the Mexican-American and the American Indian often face *because of their social and cultural backgrounds* and not as a result of their economic condition. Of the numerous problems that students encounter in deciding whether or not to pursue science, most have to be faced continuously through life. The following seven merit discussion in the order stated because each contributes to the next one.

1. First, there is a problem because much of what a Mexican-American or an American Indian *values* differs from what is valued by the society of the majority. This is, of course, something that many—teachers, other educators, and scientists—in the majority's society recognize. The differences in values of these groups by themselves would not be a serious problem were it not for the ignorance of teachers of many specific differences in values between major cultures and the cultures of the minority. The cultures of the Mexican-American and the American Indian *have been able to accommodate* changing values; for instance, the intervention of science in the healing of the sick, in transportation, in communications, etc., has been accepted in these old cultures and no longer presents a part of the dilemma. Conversely, resistance to change of some important values does present a dilemma that is illustrated a little later on in this paper.

2. The differences in values of each of these cultures from those of the majority's society do rob Mexican-Americans and American Indians of the power to determine how terms will be used in society; that is, not only does the use of the English language determine what words mean, but the values of the society of the majority also dictate how words will be used.

The principal difficulty does not arise merely from a student's inability to perceive how a teacher uses a term or how a term is valued by the teacher (or any other member of the majority). A problem surfaces when the teacher either negates the student's countervalue or, because of adherence to a stereotypical perception of the student, cannot believe that the student is capable of using a particular term in the same manner.

3. The student encounters difficulty in a culture in which the thorough mastery of nature is pursued. Intellectually, the student may consider that complete knowledge of nature is neither possible nor desirable; management of his or her environment is an alien concept. Control of nature to a degree is desirable, to heal the sick and to provide for the immediate necessities of life. However, a life in harmony with nature, and occasionally in submission to it, is not only a part of religious life for many of our people but is part of an existential philosophy that dictates that we provide for today and allow *nature* to help provide for future generations.

Clashes with the larger society are inevitable. In the classrooms, teachers request that students collect insects, dissect animals, or dig the remains of older societies. For students whose acculturation is only beginning, the destruction of life or the uncovering of the skeletal remains of humans prohibits him or her from pursuing science as it is studied in the major culture. The student's counterexperiences include familiarization with an oral history of his or her people, which is considered as valuable a record as that devised by the larger society.

4. The observation is often made that students of these minority cultures tend to value immediate rewards for study. The prevailing existential philosophical attitudes of their culture contribute to this attribute. Second, the measurement of time to conform to short-range goals and expectations is part of what makes this student different. Fundamentally, the student does not have a problem. It is the teacher or the institution of learning, which has not adapted to the student's perception, that is in difficulty.

5. The society's perception and value of pure science, in contrast to applied science, is perplexing to the minority student. The arguments concerning the worth of each are not fully comprehensible. For students imbued with a fair amount of idealism and a desire to improve their communities, it is disheartening to deal with the snobbishness of many scientists, who feel that the usefulness of their work should never be questioned by students. Students occasionally obtain the impression that they are being told that some scientific work is valued precisely because it is useless. This example, of course, can be carried to an absurd extreme; however, the perception of the student is important to him or her.

6. The minority student most often suffers from the idea that he or she cannot possibly study science because of a presumed lack of ability. The society-at-large believes that science, above all else, is the pursuit of knowledge in some very rigorous and systematic manner. Associated with this rigor and precision is something called the scientific method. And, of course, that is precisely what the minority student believes. The student has a familiarity with chance discovery, serendipity, and trial and error from his or her home and culture but acquires the distinct impression that these things have no place in science.

7. Finally, the student faces the problem of rejection by members of the major culture. The real dilemma for the minority person is that adherence to his or her own traditional beliefs earns rejection but that acculturation and the adoption of new values and modes of behavior do not ensure acceptance into the majority's society. This situation is not unusual for the minority person in any segment of this society. In that subset of society occupied by scientists, however, a few matters tend to aggravate the rejection that is felt by the minority person. A part of this rejection has to do, as we described earlier, with the way that terms are defined. For instance, thinking and discourse in science are "rigorous." The opposite of that, as the minority student perceives the definition, is the method by which members of his or her culture have gained knowledge and passed it down from generation to generation. The result is that even the method of his or her own reasoning is placed in doubt.

Another characteristic that scientists seem to possess, to the unsettlement of minority students, is a strict adherence to linear time. Nonlinear time, in the majority's culture, seems to be reserved for literary people or for people who are viewed to be not wholly in touch with reality. The student finds it necessary to shift to the accepted pattern of behavior.

Rejection of the minority person who adopts the major culture's behavior patterns can probably be explained in various ways. The minority student's analysis is, however, that he or she is rejected only insofar as the stereotypes of the older culture are applied to him or her. The net result is that the society values him or her less as a scientist or as a potential scientist.

The number of Mexican-Americans and American Indians in science will not increase appreciably with the traditional efforts: recruitment, scholarships, tutorials, and counseling. It is equally important that (1) scientists and teachers have a knowledge of minority cultures and (2) that teachers be sensitive to the students of these cultures. Sensitivity, in the psychological sense, requires that acculturation be a bilateral process between teacher and student. Compromises between cultures that are not achieved early by the mutual agreement of the parties are arrived at later very painfully.

8

Problems of Minorities at Majority Institutions: A Student's Perspective

WOODROW A. MYERS

The hardest burden that a minority student at a majority institution has to bear is the alienation felt when one realizes that few share one's point of view. It is not the academic competition that one faces, since one also competes with the brightest students in the country when enrolled at Howard or Meharry; rather it is that of being different and misunderstood It is that self-esteem and self-respect are repeatedly challenged on a racial basis, as well as on a personal basis. There is repeated challenge in all conceivable situations from the first day of med school, when your white lab partner wonders aloud whether one has to keep the same partner throughout the year, until graduation when one sees fewer black classmates than were there on the first day. The challenge is internal as well as external. One is forced into a long "Are you who you think you are?" period and there is little help around to see you through. Many times in the process you want to give up and say that you tried, but more often, fortunately, you try harder and keep going.

The problem of assimilation into an organized system of higher learning is not the problem solely of minorities at majority institutions, but the problem of many students at all institutions. The problems that are peculiar to minorities at majority institutions must not be confused with the problems of minorities in changing academic settings, the problems of minorities in adjusting to medical-school curricula, or the problems of minorities in attaining academic excellence. These are problems encountered by all students. Why, then, is there such a statistical difference in what happens to whites and nonwhites as they enter professional education?

WOODROW A. MYERS • Harvard University School of Medicine, Boston, Massachusetts.

According to figures released by the Association of American Medical Colleges (AAMC), 1,036 black students entered the first year of medical school in the beginning of the 1975–1976 school year, making 6.8% of the total first-year class in 114 medical schools across the United States (AAMC, 1974, 1975).* The year before, in the fall of 1974, 1,106 black students entered the first year of medical school, making up 7.5% of the total. Therefore, in one year, the number of black medical students entering medical school dropped by 70, while the number of white students entering medical school increased by 614. Discrimination is not only very much in drive; never having been in reverse, it has shifted into high gear. This is the first year that the number of black first-year medical students has decreased since a generally active recruitment was initiated in 1968. There were also similar decreases this year in American Indian and Mexican-American admissions.

Most people around you do not realize that no one gets into medical school because of being black but rather in spite of it. The prevalent attitude among white medical students is that minority students were admitted under lower academic standards, in other words they (white students) worked harder to get to the same place as the minority students. Most minority students were admitted through special programs, programs that for the most part allowed flexibility in rigid, strictly academic admission criteria: programs that looked for people who demonstrated excellence in a wide range of academic and extracurricular activity; programs that were designed to begin to desegregate rather than discriminate. All too frequently minority students find themselves as defendants. We are always called upon to give a minority point of view, whether we have one or not.

The December 1975 issue of the *Journal of the American Medical Association* was almost entirely devoted to the Council on Medical Education of the AMA's 75th Annual Report on Medical Education in the United States. Of the 107 printed pages, 1⅙ pages are devoted to students of minority groups. We find on these pages that 14.4% of the black students in the 1974–1975 entering medical-school class repeated the first year. The actual number was 161 out of 1,117. There were 162 nonminority students in the same class repeating their first year, that is, 162 out of 13,472. Even in the second, third and fourth years, black students repeat at a rate that is 10 times that of other students. Further-

*See the tables included in the article by Maxine Bleich in this volume.

more, of all the black students admitted in the fall of 1972, 13% were no longer in medical school in June of 1975. The attrition rate for nonminority students during this period was 2%.

If the transition from undergraduate school to professional school is no different for minority students, and the adjustment to the workload in medical school is not genetically correlated, and the goal of achieving academic excellence is sought by all students, then why are there so many discrepancies between what happens to minority versus majority students after they enter professional school?

I (and most of my colleagues) have found that racism takes a thousand shapes in the academic setting. Let us take a look at a conference with Professor A, who gives his monthly lecture to medical students electing the course relating to a specialty. Many times, before the lecture or the conference begins, the instructor initiates informal chitchat with individuals in the group. In these small groups, there are at most two minority students, and when the conversation comes to me, as it invariably does, I find myself giving the same answers to predictable questions. The answers usually go something like: (1) "No, I go to this medical school; I'm in the same class as most of the students here." (2) "No, I did my undergraduate work at a fairly large school." (3) "Actually my B.S. was in biology." (4) "No, I grew up in the North, so I'm used to cold weather and snow." (5) "No, I didn't play football, basketball, or throw the shotput in college; I found myself too busy with biology, chemistry, and the like." (6) "Yes, I do find the work here difficult, but it's something I'm used to by now." (7) "No, I haven't decided on community medicine or primary-care practice." (8) "No, actually there are no physicians in my family, and no one is in any way connected with medicine or with a medical school." When the conference finally starts, one of the hardest things to do after this introduction is to convince yourself that there is no reason to feel paranoid. You tell yourself that if asked a question, you will respond as best as you can. But when the question comes and the answer is not at the tip of your tongue, you sometimes feel as if again you have been singled out. In your anger, acute brain–mouth block sets in, and the answer that you ordinarily could have put together verbally remains locked in your mind, right beside the realization that you performed as "they" expected you to perform.

Then there is Professor B, who makes a special effort to befriend the minority student. He tells you that he realizes that you may have "special difficulties" and that his office is open at any time for you to come in and discuss things. Although such a gesture on the surface may seem friendly and innocent enough, when one goes to Professor B's office one finds oneself denying questions about hostilities he assumes you have toward white interns, residents, or fellows and denying the kind of difficulties he assumes you have in such a highly academic environment. When your chance comes to discuss the question you had in mind, your level of knowledge is assumed (as deduced from the explanation given) to be somewhere in the range of a sophomore premedical student.

Professor C thinks that it is wonderful that so many of "you people" are finally getting a chance and tends to constantly let you know exactly that. He may feel that you are a genius, since you are black and "made it" in a system so geared against you. This is a theory that he frequently tests with the questions he asks you versus those asked of the other students.

These are blatant examples of discriminatory attitudes toward minority students and are encountered only infrequently as against the more subtle forms that are a day-to-day occurrence. The subtle forms occur, for example, in statements made regarding "this type of patient," or "that socioeconomic class," or "in this geographic region." In discussions ranging from mental retardation to cervical cancer to *Hemophilus influenzae* meningitis, the medical student is constantly made aware of racial and socioeconomic differences within the spectrum of a disease process. Interestingly and unfortunately, the discussion of these differences in disease patterns rarely center on the reason for their continued existence. Too frequently, minority students feel as if they are the only ones who see the hypocrisy and the callousness of the academic setting. Too frequently, they ask "Why?"—only to be batted down by "That's the way things are," or "Perhaps your generation of physicians can do a better job," which seems highly unlikely when many times we are taught pathophysiology to the exclusion of pathogenesis and prevention.

The minority student at the majority institution often gets the overwhelming feeling of "aloneness." There are very few of us in any one course at any one time. Perhaps there is one other. Outside of coursework, relationships between minority students are frequently not all that they could be. Fortunately, there are more of us around now than there were 5, 10, or 20 years ago, but we do not take full advantage of the situation because of the strong assimilatory pressures of the institutional environment. There is, however, an organization called the Student National Medical Association, which is predominantly black and has local chapters at each school with minority students. Not only does this organization provide the opportunity to come together and talk about issues of importance, but it also allows an opportunity for socialization, a rare event in the already-overscheduled day. Too frequently, many minority students think that this is time better spent with the books, believing that time spent away from studying will add nothing to the grade-point average. More often than not, it is these students who find life in medical school more difficult.

The sense of "aloneness" is reinforced by racism as it exists in the larger medical community. For example, while sitting in the lobby of one of our teaching hospitals, I counted the first 15 black males around my age who passed by, noting their position within the hospital: 4 were janitors, 4 were food-service workers, 4 were patient escorts, 2 were visiting patients, and 1 was a respiratory-therapy student. However, of the first 15 white males of similar age, 5 were medical students or interns, 3 had administrative positions, 2 were policemen, 2 X-ray technicians, 1 patient escort, 1 hospital volunteer, and 1 was a patient. Thinking about the reasons for such differences keeps one from getting caught up in one's own successes because one realizes that very little has changed for the vast majority of black people.

You try to be congenial with black hospital employees, yet you find too often that a white coat and a stethoscope turn a bold, loud "What's happening, brother?" into a mumbled hello or a no response. Black instructors and professors are, of course, rare or absent at Harvard as at most majority medical institutions. I encountered my first black instructor well into my third year of medical school, after almost all of my major clerkships were completed. This all means that the role models one uses are white, make-believe, or nonexistent. Too many black medical students lose black identity, lose a positive black self-image, lose all contact with the black community, and become immersed in and an appendage of, not a part of, the white medical culture. One finds oneself becoming more able to tell the difference between dry and cream sherry but less able to sympathize, less willing to empathize with the 34-year-old alcoholic on his or her 10th hospital admission.

The pressure on minority students is not to integrate but to assimilate—not just the factual and theoretical medical information, but in addition, to assimilate into a way of life that is alien, that represents much of what you have been denied, that represents the wrong side of the social-class system you claim to reject and a way of life in which you find little support for the idealistic goals that you rapped about in your medical-school interview. These were goals that you naïvely thought you could readily accomplish.

The attrition rate for black students is higher than that for whites. In thinking about what we minority students go through, it does not surprise me. Every black medical student I have known who has left medical school has left under intense emotional stress, not on a whim to travel around the world, or to try out business school instead of medicine. The stress began long before the time for the biochemistry final or the national boards. These exams were only the bullets shot through the pistol-whipped mind. Why do others make it through? Perhaps because some of us band together in socially active units, at least listening to Grover Washington and Gladys Knight, even if we do not have time to go to see them. We come together knowing we should drink less than we used to but find ourselves drinking more. Some of us "go it alone," becoming studying and test-taking machines, literally spending four years moving from dorm room to classroom or hospital floor and back again. These individuals are capable of successfully suppressing the frustrations. They are few in number, and occasionally their methods of dealing with reality tragically fail them. For those who have no methods, though, coping is impossible and they become the casualties.

For the last five years, I have been asked to speak at conferences attempting to deal with this same topic, conferences that by their very nature attract those who are sensitive to and already interested in this topic. Those who need to become aware of the pressures leading to the problems—the admissions committees, the chairmen and professors of clinical and preclinical sciences, and the house staff with whom you spend much of your time—are the ones less likely to be present at these meetings. We will not leave this assembly with many answers, but hopefully we will be better able to ask the questions concerning the pressures faced by minority students. As minority house staff become more numerous, the number of role models will increase, and as some of them become professors and their sensitivities (hopefully) rub off on their colleagues, the frustrations we students now face will not be those faced by our "brothers" and "sisters" five years from now. It is a very hard thing to wait for these changes, and harder still to believe they will happen.

The first time I went to see a professor in medical school on an informal basis, without a white coat and a tie, I remember his secretary looked up and said before I could utter a word, "I didn't know we had any deliveries today." From that time onward, I learned that it is much more effective to put a title behind your name and to look the part and act the part, even when introducing yourself or making an appointment. Today, I wear a white coat when the image it creates is useful. The purpose of medical school, in a way, is to learn what is expected of those who hold the title, to help those who respect it, and to learn to use the privileges of those who have it on. The trick for the minority student is to know when to put it on, but more importantly to know when to take it off.

REFERENCES

Association of American Medical Colleges, Division of Students Studies. *First-year U.S. minority and foreign student enrollments,* 1974.

Association of American Medical Colleges. *Fall enrollment questionnaire: Students in 114 U.S. medical schools,* 1975.

Journal of the American Medical Association, 1975, *234,* 1339.

9

Medical Education and the Minority Student

STANFORD A. ROMAN

Any discussion about medical education and minorities must consider the weakness of the medical-education process as it impacts on all students. While the skills of the surgeon may be distinct from those of the internist, both skills are superimposed on a general base that assumes that students should be well versed in all the basic sciences. Since the report of the Carnegie Commission in 1910 (more popularly known as the Flexner Report) prescribed that medical education should include training in the basic sciences (e.g., biochemistry, anatomy, and microbiology) and clinical preceptorships, the medical curriculum has changed very little. Although student activism in the late 1960s led many schools to increase the number of optional courses in the last two years of training and to include community medicine as an area of speciality, these changes have had a minimal impact on medical education *per se* and thereby on the product of that education, that is, the practicing physician.

Upon admission to medical school, the student is exposed to a curriculum that dates back to 1910 and is proclaimed to be associated with the training of a "good physician." Each department is allowed to assess its resources and develop its course content independent of its direct relevance to human health. There is a conflict here: while certain courses have been defined to be associated with the training of a good physician, there are few guidelines to determine the importance of each course to the practice of medicine. Hence, the student must absorb the favored project of the scientist studying the venule structure of the

STANFORD A. ROMAN • Assistant Dean, Boston University School of Medicine, Boston, Massachusetts.

tsetse fly as energetically as that of the cardiologist who is studying enzyme transfer in the myocardial muscle of the dog.

The medical-school faculty member most commonly is a white male whose family orientation is upper middle class. He is involved in research and probably devotes little of his time directly to patient care. He knows that if this balance is lost (i.e., too much time devoted to patient care), he may be relegated to the "dubious" category of clinical faculty. It is amusing that while most physicians on medical-school faculties regard the clinician with respect, the designation of clinical assistant professor or assistant clinical professor is regarded as second-class membership in the academic fraternity.

The student is soon socialized into these hierarchical patterns. Most students look toward their teachers as role models; it is then not surprising that medical schools produce students after the image of their faculty. The student who is able to conform to and succeed in reflecting this image, both academically and socially, will be rewarded with a university hospital internship (the most prestigious of internships).

The medical educational system of this country produces some of the finest technocrats in the world, individuals who can respond with competence to most medical diagnoses but who, in moments of frustration, view a run-of-the-mill patient as "a turkey" or "a crock." A patient is seen simply as a collection of parts, and the humanity is lost to "the appendectomy," "the herniorrhaphy," "the amputation," and "the hemipelvectomy."

In 1969, the Association of American Medical Colleges (AAMC) recognized that minorities were underrepresented in the medical profession: minorities defined as blacks, mainland Puerto Ricans, American Indians, and Chicanos (AAMC, 1969). While we commend this recognition, we must ask what has happened since 1969. In 1970, the task force suggested that 12% of the entering class in 1975–1976 should be comprised of blacks (Nelson, Bird, and Rogers, 1970, 1971). In reality, 6.8% of the entering class of 1975–1976 were black, which represented a decline from the preceding year, when 7.2% of the class was black.

The assumption in 1970 was that a short-term effort in recruitment and retention could satisfy the enrollment goal of 12% blacks by 1975. Even though social and economic disadvantages were found to be associated with less than traditionally accepted levels of performance in the undergraduate phase, the impact of the deprivation was not fully recognized. In the case of many minority students, the precollege experience has probably not provided the quantitative and qualitative skills necessary to meet the challenge of a rigorous premedical curriculum. The maintenance of a strong science curriculum on the high-school and junior-high-school level is expensive. Economic disadvantage makes the road to medical school difficult for the poor and the minority student at an early age. During high school, the minority student who may already experience reading difficulties or inadequate study skills is discouraged from academic pursuits, especially those leading to a health profession. Most often, accolades go only to the student who aspires toward professional sports. While the ticket out of the ghetto provided to many minorities by professional sports is important, we must be concerned about a system that largely encourages the minority student to attend college with the major aim of entering the professional athletic pool. In college, such a student is less likely to undertake the time-consuming premedical curriculum. It is therefore not surprising that the 1970 projection—6% of minority college freshmen specifying medicine as a career choice—has not been realized (Nelson *et al.*, 1970).

The minority student enters college with long-time inadequacies in earlier educational experience. Often, the need for academic remediation prevents immediate entry into the premedical track. Second, the disadvantaged student, in general, has had little exposure to the health-care system or to health professionals. The student may choose or not choose a health career simply because of lack of information. Decisions may be made based on the length of time required in school, the ultimate economic rewards, and the number of laboratories required per week. While all these factors require consideration by the student, the isolation of the minority student from the health professions reduces his chances of making decisions based on long-term advantages as against short-term benefits.

In the undergraduate college, the minority premedical student is often isolated, particularly in the predominantly white university. This student becomes a minority among minority students. The traditional premedical-adviser system within colleges is often not supportive of the minority student who does not manifest performance levels comparable to those of the majority student. I have concerns that these systems, in fact, may result in discouraging even the competitive minority student. The student, therefore, finds no incentive for traditional programs. In recent years, minority students have organized premedical clubs and societies in an attempt to come together as a group and to provide pertinent information about medicine and medical school. The effectiveness of these clubs has varied, as have their numbers.

The student who does survive and gains admission into the medical school is viewed as being different. The minority student is by and large not upper middle class. The experiences that he brings with him may be different from those of his mostly white classmates and predominantly white teachers. His motivation to enter medicine may be less "sophisticated" than that of his white classmate. How then does this student adjust to an environment that is basically white and upper middle class and that is more often than not resistant to change? The anxieties inherent in such a meeting are evident. Medical schools, through minority-affairs programs, attempt to "help the student adjust and progress through the curriculum." Unlike his white counterpart, the minority student finds difficulty in seeing the typical medical-school faculty member in a predominantly white institution as a role model. Minority faculty members are rare. If, indeed, the school attempts to involve practicing minority physicians, the academic elite regard them as second class. Within the medical school and the university, the practicing physician is by definition an outsider, a local medical doctor (LMD). If he is involved directly in the medical school, he is labeled with creative academic titles such as "adjunct professor" or "clinical associate."

In medical school, the inclusion of the minority student is accepted to differing degrees by other students and faculty. The DeFunis Case (*DeFunis v. Odegaard,* 82 Wash. 2d 11, 507 p. 2d 1169, 1973, 416 U.S. 312, 1974) has provided a reason for medical-school faculties to mumble about "letting in all these 'poor' students." Ironically, these "poor" students require good teaching, a phenomenon frequently not essential to medical education. As medical-school admission becomes more competitive, the quality of the majority student, as measured by traditional performance criteria, has improved. The inclusion of the minority student who has good potential but not the quantitative and qualitative skills possessed by most majority students may require teaching skills that are not traditional to most medical school faculties. It is ironic that despite the stated objectives of teaching, most medical-school faculty are primarily researchers and only teachers secondarily. The academic faculty, therefore, responds to teaching demands with varying emotions of frustration, hostility, and despair. Many students, and among them most minority students, find such an environment alien and unfriendly.

The underrepresentation of minorities in medicine is more than a black, brown, and white issue. While it is naïve to believe that because one is a member of a minority, one will automatically serve in a minority community, it is equally naïve to believe that health services within inner-city and rural areas can be substantially increased in the absence of a significant increase of minority physicians. The issue must not be mere representation but representation for what purpose. A national health-service corps with available financial aid can provide short-term aid to these areas, but we must be cautious about such arrangements. Our goal cannot be to provide underserved communities with a different physician every two, three, or four years, while the middle-class communities are assured of continued and uninterrupted medical services. We can no more expect a significant number of white middle-class physicians to practice in the Harlems of our country for a long term than we can expect a significant number of black physicians to practice in all-white communities. While race and ethnicity are not the sole determinants of practice site, their role cannot be underestimated.

My despair rests with the fact that we are yet to examine medical education in a comprehensive and critical fashion. Family practice and primary care have recently been established as separate specialties. Can the concept of primary care be effectively taught in a domain where the subspecialist and the specialized researcher reign supreme? Can the present system that has demonstrated a capability for training highly sophisticated physicians, in a technical sense, also be capable of imparting the deep sense of humanity and the awareness of the patient as a human being (rather than as a collection of parts) that is required of a good physician, especially a primary-care physician? Can such changes in attitudes and purpose come about on the part of the medical system without its undergoing an intensive and critical self-examination?

In conclusion, if we are to address the underrepresentation of minority students in medicine, we must wage our attack on several fronts. First, the elementary and high schools must develop adequate reading and study skills as well as adequate quantitative and qualitative skills in their students so as to enable them to handle the college curriculum competitively. Second, the undergraduate colleges, which have increased their enrollment of minority students, must develop the necessary academic supports required to prepare these students for graduate study. The identification of potentially able students who, because of early educational disadvantage, may require additional preparation in reading, study skills, and communication has not received much attention. Third, colleges must develop on-site career experiences that will give minority students several viable alternatives before they make their career choice. Fourth, premedical advisers should advise students in a fashion that will allow them to make the right career choices, giving ample consideration to the able student who, with appropriate support, may be able to complete medical school successfully.

The medical school, through its selection of students and faculty and its economic and status dependency on research, turns out a technically competent physician who will probably obtain subspecialty training and will probably locate in some place other than an underserved community. It is sophistry on the one hand to stress clinical services and on the other hand to provide as major role models a group of subspecialized researchers. I am not professing the destruction of the present medical educational system but simply advocating a reassessment of its objectives. When an educational system is unresponsive and insensitive to the needs of society and becomes very much insulated and isolated in its functions, then it is the responsibility of the public to restore that educational system to the serving of the society. The increased representation of minorities in medicine is only one indicator of the commitment to equal access to health services, not only of minority and poor communities but of the nation as a whole. If the nation is indeed planning for an equitable national health-care system, then the concerns discussed in this paper become important considerations.

REFERENCES

Association of American Medical Colleges. *Proceedings for 1968. Journal of Medical Education,* 1969, *44,* 349–469.

Nelson, B. W., Bird, R. A., and Rogers, G. M. Expanding educational opportunities in medicine for blacks and other minority students. *Journal of Medical Education,* 1970, *45,* 731–736.

Nelson, B. W., Bird, R. A., and Rogers, G. M. Educational pathway analysis for the study of minority representation in medical school. *Journal of Medical Education,* 1971, *46,* 745–749.

III. Public Policy and Biomedical and Behavioral Training: Effective Development of Existing Potential

10

Public Policy for Minority Self-Actualization: Present Realities and Future Possibilities

VIJAYA L. MELNICK

I am privileged to introduce to you a highly talented and articulate panel. My esteemed colleagues will address specific policies and their impact on minorities in the area of biomedical and behavioral sciences. The effect of such policies on higher education will be discussed. I shall use what time I have to discuss the environment and the general milieu within which these policies ought to be generated, examined, and implemented and to suggest some strategies for consideration in the generation of new policies that affect the training of minorities in the health sciences.

It is an inescapable fact that any question that addresses minority participation, be it in science or economics, politics or humanities, is inextricably tied to the social phenomena of this society. It is tied to the historical events of the last 200 years that shaped this nation. How successful have we been in translating the lofty ideals of the American Revolution to all of society?

Equality of opportunity is one of the great boasts of American democracy. American folklore is replete with stories of those who "made it": the homespun heroes who went from rags to riches, from log cabin to the White House, from illiteracy to intellectual scholarship. Many Americans, however, have stood outside the pale of the "American dream"—for there have been grievous inequalities in this modern cradle of democracy. Abraham Lincoln was perhaps more realistic in assessing the idealism of the Declaration of Independence: he noted

VIJAYA L. MELNICK • Associate Professor of Biology, Federal City College, Washington, D.C.

that "It meant to set-up a standard maxim for free society which could be familiar to all, revered by all. Constantly looked to, constantly labored for and even though never perfectly attained, constantly approximated, and thereby constantly spreading and deepening its influence, and augmenting the happiness and value of life to all people of all colours everywhere."

How has this process worked since the emancipation? During the times of Franklin Roosevelt, John Kennedy, and Lyndon Johnson, public programs to redress the grievances of the poor, the minorities, and women were introduced. Some brought forth radical social change; others were slow and plodding. In spite of these measures, the plight of the majority of black, native, and brown Americans remained a glaring contradiction to the American claim of equality of opportunity.

It was only 22 years ago, on May 17, 1954, that the Supreme Court of the United States unanimously ruled that *"separate cannot be equal."* Until then, the Supreme Court ruling of 1896 *(Plessy v. Ferguson)* had upheld, blessed, and legalized segregation. It was this decision that condemned generations of blacks and other minority children to dilapidated schools, outdated books, outmoded education, and closed horizons of expectation. Even such an education had to be attained at a tremendous cost. It involved walking long distance (not riding in buses, mind you) to segregated schools at great personal risk and hazard.

Richard Kluger (1976), in his recent book *Simple Justice,* eloquently recounted the very-low-profile insurrection of 1947 in Clarendon County, South Carolina, led by the Reverend DeLaine to protest the deplorable educational conditions of black children. As a result, DeLaine unleashed upon himself and his family a reign of terror. Kluger recounts how before it was over, they had fired DeLaine from the little house at which he had taught devotedly for ten years, burned his house to the ground, and stoned the church which he pastored.

DeLaine's unpretentious protest, however, was to be the primary force that gathered together many dedicated people whose efforts eventually led the way to the momentous 1954 Supreme Court decision.

According to Kluger, nothing much has changed in Clarendon County. There, integration is yet to come. There are then places, including some large urban areas, in these United States where "equality of opportunity" is only a phrase in a historical document, for the concept still remains alien to the environment and to day-to-day life.

The eminent sociologist Kenneth Clark (1967) incisively wrote, "The dark ghettos are social, political, educational, and—above all—economic colonies. Their inhabitants are subject peoples, victims of greed, cruelty, insensitivity, guilt, and fear of their masters" (p. 11).

The economic deprivation of the large numbers of minority Americans undercuts almost all questions and poses as the major reality to be encountered when any policy that has to do with opportunities for minorities is sincerely explored.

It is of interest here to recall the 1969 memorandum written by Daniel P. Moynihan (our recent flamboyant ambassador to the UN) to President Nixon on the "position of Negroes" (Moynihan 1970a,b). He wrote, "The Negro lower class must be dissolved by transforming it into a stable working class population." It is "the low income marginally employed, poorly educated, disorganized slum dwellers whom Black extremists use, to threaten white society with the prospects of mass arson and pillage." Moynihan then went on to offer the remedy of "turning these people [i.e., the ghetto blacks] into truck drivers, mail carriers, assembly line workers—people with dignity, purpose, and in the United States, a very good standard of living indeed." Comparing favorably the income attained outside the South by young black families and young white families, he concluded, "The time may have come when the issue of race could benefit from a period of 'benign neglect.' The subject is too much talked about. . . . We may need a period in which Negro progress continues and racial rhetoric fades" (quotations are from Moynihan, 1970a, p. 69).

I want to remind you, my dear friends, that we are still in the period of an administration (even though the central characters have changed) that made this memorandum their manifesto and major reference point to deal with the minority plight.

Andrew Brimmer in his speech at Tuskegee (1970) warned "that it would be a serious mistake to conclude that the Black community has been so blessed with the benefits of economic advancement that public policy which played such a vital role in the 1960's, need no longer treat poverty and deprivation among such a large segment of society as a matter of national concern. To accept such a view would certainly amount to neglect—but it would be far from benign."

The trends of the last five years have proved those words clairvoyant and the neglect not benign but cancerous. A recent report on labor (Greenberg, 1976) by the National Association for the Advancement of

Colored People (NAACP) reveals that during 1969–1970, black income had reached 61% of the income earned by whites. By 1972, it had dropped to 59% and by 1975 to 56%. This figure is lower than that during the period following World War II, at which time the black family's income was 57% of the white family's. In explaining the grim picture, Herbert Hill, National Labor Director of the NAACP, observed that discrimination in employment is not the result of random acts of malevolence; it does not usually occur because of individual bigotry, but rather in the consequence of systematic institutionalized patterns that are rooted in the society. Thus sweeping changes are necessary if racial employment patterns are to be fundamentally changed.

The unemployment rates illustrate the point. Official reports showed a rate of 7.6% for whites and 14.1% for blacks in October 1900. The National Urban League and the NAACP pointed out that if "hidden unemployment" had been fully considered, then the rate was close to 13.6% for whites and 25.5% for blacks. Moreover, in the major areas of black urban population, black unemployment reached a whopping 30% for adults and over 40% for teen-agers.

The deplorable economic status correlates in turn with depressing health status. The recent report on "Health, United States 1975" showed U.S. infant mortality rates at 16.5 per 1,000 births, but the rate is 27.5 for black babies. Similarly, the life expectancy of black males and females is well below that of their white counterparts. (Infant mortality and life expectancy are generally regarded as the two areas that indicate general health status.)

On the health professional level in 1969, 2.2% of the physicians, 2% of the dentists, and less than 5% of the nurses were black. In 1975, the percentage of black physicians actually dropped to 1.9%. The number of black freshmen in American medical schools had increased from 4.2% in 1969 to 7.2% in 1974. In 1975–1976, this figure dropped to 6.9% (Association of American Medical Colleges, 1973, 1974). Dr. Max Seham (1973), a perceptive and concerned white physician, observed that the victims of racism within our medical system have not been limited to the patients. Discrimination and poverty have also had a disastrous effect on the opportunities for black medical students. The obstacles that stand in the way of the potential black American medical student and physician are many. The most critical of these being the existence of covert and overt racism and sexism, a handicapped educational background, the economic problems caused by the escalating cost of medical education and the dwindling sources of support, a lack of total social and professional acceptance, and the relatively low numbers of black doctors at decision- and policy-making levels in professional institutions and organizations. Much of the same can be said for black undergraduate and graduate programs in basic biomedical sciences. The percentage of black Americans with Ph.D.'s in basic biomedical science, for example, is only about 1% (see James M. Jay elsewhere in this volume).

If we do not find many blacks and other minorities in health professional schools and graduate programs, in what area of the biomedical sciences *do* we find them?

The place where they are found in overwhelmingly large and disproportionate numbers is in biomedical human experimentation. A conference on this subject in 1975 called attention to the fact that minority groups and the poor were consistently exploited in biomedical tests and experimentation. Speaking at the conference, Carl Holman, President of the National Urban Coalition, said, "We don't want to kill science, but we don't want science to kill, mangle and abuse us."

It is a matter of great shame and tragedy that many scientists have been implicated in making use of powerless groups, such as the poor, the disadvantaged, and minorities, for biomedical experimentation in the name of benefit to the society at large.

One has to question the ethics of such procedures, the morals of such behavior, and the cost of these benefits to society. Is it not time that we ask what science can do to redress the tragic condition of the poor and the disadvantaged rather than continuing to ask what the poor and the disadvantaged can do for science?

As scientists and educators, we find ourselves largely ill informed and apathetic, our hands stained with our brethren's blood; as citizens, we act as though we are in the grips of a galloping epidemic of the ostrich syndrome. It is time to wake up from the sedation of a few illusory gains. If we do not make every effort to work for policies, programs, and dollar commitments that will counteract these dangerous trends, the American dream will surely become the American travesty.

In words of Frederick Douglass (1857), "The whole history of the progress of human liberty shows that all concessions yet made to her august claims, have been born of earnest struggle. . . . If there is no struggle there is no progress. Those who profess to favor freedom and yet deprecate agitation are men who want crops without plowing the ground, they want rain without thunder and lightning. They want the ocean without the awful roar of its many waters. . . . Power concedes nothing without a demand. It never did and it never will."

The absence of any mention of civil rights, the plight of the minorities, and progress toward achieving equal opportunities is to be noted in both the President's 1976 State of the Union Message and the Democratic reply. Their silence is thunderous. As Richard Goodwin (1974) pointed out, basic themes of oppression have never been made in grand political statements. They can be recognized only as hints supplied by subtle changes in the manner of presentation and the character of public response. Problems of poverty, unemployment, civil liberties, and income distribution have been around for a very long time and continue to be. What is distinctive is the passion and the commitment with which these are debated from time to time. That difference measures the order of priorities of an administration or a government.

Contrast, for instance, the present silence on such issues to the not-so-long-ago time when a President of the United States could affirm the sentiment of a forgotten people and proclaim "we shall overcome" to almost universal applause. Needless to say that impassioned public rhetoric and the laws that were enacted at that time should be followed by strong and supportive action. If not, no changes need be expected in the dominant institutions of our society. Without such change there can be no progress.

We must, on our part, strive to put forth all effort to ensure such a commitment and action from those that make and implement policy.

In the work for future change and progress in the biomedical arena, I *underscore* the following strategies:

1. Major decision-making bodies that set priorities and formulate policies for medical training and services should include consumer representation, for medicine is not solely a technical matter. It has an inherent and significant societal component. This aspect has largely been ignored up until very recently. The consumer voices should come from all walks of life and represent the demographic patterns of a given area. The need for such participation has already been shown by studies of Mechanic (1974), Etzioni (1968), and others.

2. A major commitment should be made by policy and financial allocation to strengthen the undergraduate and possibly the preundergraduate training of minority students. Significant numbers of minority students come from small colleges that are underfinanced and understaffed. Thus, they lack the resources for setting up programs in modern molecular and quantitative biology that require commitments of not only dollars but time, space, and personnel. Innovative programs, however, need not only imagination and enthusiasm but cold, hard cash.
3. There should be encouragement and expansion of summer/special programs for minority students from small colleges who are interested in careers in biomedicine. Such programs should be located in university medical centers or large research laboratories. The experience will not only help reduce the trauma of transition for the students but give the universities and the laboratories a chance to understand and appreciate some of the real problems that confront the students. A mutual education is both advantageous and necessary.
4. If significant increases are to be effected in the production of minority health professionals, two simultaneous strategies seem warrented. (a) Some of the new biomedical centers should be focused on a largely minority clientele. The need for such an approach is eloquently made by Dr. Therman Evans (1976), Executive Director of the Health Manpower Development Corporation, Washington, D.C. He pointed out that over the last five years, the total number of medical students increased by close to 18,000. In 1975–1976 alone, when there was an increase of over 500 places for entering medical-school freshmen, the number of black medical-school freshmen decreased by over 80 from the previous year (1974–1975). (b) The opening of new facilities for training largely minority medical students should not be taken to mean that other medical schools can abdicate their responsibility to recruit and train minority students. The gap between the need for minority health professionals and their existing numbers is so great that only a multipronged effort will lead to closing that gap.
5. The formation of an interested and dedicated group of biomedical scientists is needed who can act as a liaison between policy

makers (such as the appropriate offices of the executive branch and the committees of the Congress) not only to help evaluate and comment on current biomedical policies but to suggest new policies and legislation in the area of minority biomedical participation. These efforts should be made in concert with such civil-rights organizations as the NAACP and the Urban League; our national policies and the government itself can become more representative by such efforts.

6. A national information-resource center should be established to facilitate the compilation of information and the discussion of special programs that are primarily designed to increase the enrollment and retention of minority students in the various biomedical-science programs. This could be an organization such as the Health Manpower Development Corporation of Washington, D.C., which has already established a computer data bank for a health-careers information system.

As we come together to consider policies and to work toward effecting change, let it not be forgotten that the ability to end poverty—to end inequities—is well within the grasp of this nation. But it *requires massive changes in the material structures and relationships that dominate present society*. It requires a reordering of priorities and a strong and committed leadership with far-reaching vision. The rising tide lifts all the boats; the uplift of the poor and the disadvantaged will infuse the whole society with a new vitality and a sense of social purpose, something grossly missing in today's society.

As we embark on our third century, let it not be said that we are a nation that won its independence only to become more enslaved in narrow, repressive institutions; let it not be said that we created immense wealth only to foster deep pockets of poverty and urban malaise; let it not be said that we have come to ride at the helm of scientific and technological progress only to become ignorant of and unconcerned about the biased distribution of that knowledge and its fruits through inequitable education, opportunity, and services; and most of all, let it not be said that we are a nation that proclaimed the "inalienable right to life, liberty, and the pursuit of happiness" only to practice unequal justice and to perpetuate prejudice.

In the final analysis, as Robertson observed, "It is not the talents we possess so much as the use we make of them that counts in the progress of the world" (Beveridge, 1957, p. 186).

REFERENCES

Beveridge, W. I. B. *The art of scientific investigation.* New York: Random House, 1957.

Brimmer, A. F. Economic progress of Negroes in the United States: The deepening schism. Founders day convocation speech at Tuskegee Institute, March 22, 1970.

Clark, K. Dark ghetto: Dilemma of social power. New York: Harper Torch Books, 1967.

Department of Health, Education, and Welfare. *Health, United States, 1975.* U.S. Government Printing Office, Washington, D.C., 1975.

Douglass, F. West Indian emancipation speech, 1857.

Etzioni, Amitai. 1968. The active society. Free Press, New York.

Evans, Therman. 1976. *The Washington Post,* Jan. 28, 1976.

Goodwin, R. *The American condition.* New York: Doubleday Company, 1974.

Greenberg, J. Affirmative action: Quotas and merit. *New York Times,* February 7, 1976.

Kluger, R. *Simple justice.* New York: Alfred A. Knopf, 1976.

Mechanic, D. *Politics, medicine and social science.* New York: Wiley, 1974.

Moynihan, D. P. 1969 Moynihan memo to President urged jobs for Negroes. *New York Times,* March 1, 1970a, p. 1.

Moynihan, D. P. 1970b. Text of memorandum for the President on the position of Negroes. *New York Times,* March 1, 1970b, p. 69.

Seham, M. *Blacks and American medical care.* Minneapolis: University of Minnesota Press, 1973.

11

Some General Proposals for Increasing the Production of Minority Professionals in the Basic Sciences

JOSEPH W. WATSON

On November 4, 1975, approximately 100 presidents of black colleges and universities met with Health, Education, and Welfare Secretary F. David Mathews to propose the establishment of a national plan to achieve parity of black and other minority representation in all areas of higher education and in all professional and technical fields by the year 2000. These college presidents were, in part, led to make this proposal out of consideration of the faltering progress being made toward eliminating the historic gaps between white and minority family incomes as indicated in Table 1 (Padulo, 1974, p. 147).

As the American economy becomes increasingly characterized by a scarcity of cheap resources, it inevitably becomes more complex and dependent on science and technology. It is therefore difficult to conceive how the historic gaps between minority and white family incomes can be reduced without a concurrent equalization of minority participation in the scientific and technical professions.

Although the achievement of parity in ethnic representation in scientific and technical fields by the year 2000 is a noble objective that many can support, its achievement is most difficult because it requires not only an immediate national commitment but also one that must be sustained for at least two decades. For example, if by some miracle the educational system were suddenly revamped so that all entering seventh-graders had equal opportunities and prospects for receiving doctoral degrees without regard to race, the population of new doctorates

JOSEPH W. WATSON • Provost, Third College, University of California at San Diego, La Jolla, California.

Table 1. 1972 Median Family Incomes

Whites	$11,549
Chicanos	$ 7,908
Puerto Ricans	$ 7,613
Blacks	$ 7,106

would not be balanced until 1991 (these students would enter high school in 1979, college in 1982, and graduate school in 1986, and they would receive doctoral degrees five years later).

Approximately 16% of the United States population is minority (Padulo, 1974, pp. 2–4). A survey by El-Khawas and Kinzer (1974) of graduate institutions indicated that only 7% of all graduate students were minority (Table 2). In the basic sciences and engineering, blacks constituted less than 2% of the graduate enrollments in these fields, although they represent 11.1% of the total population and 4.4% of the total graduate enrollment. Thus, in the basic sciences, engineering, and the most quantitative of the social sciences, economics, blacks are underrepresented by from 6% to 9%, while for all fields, the underrepresentation is approximately 2.5%. It is noteworthy that in 1973 almost half (43%) of all black graduate students were in education, compared to 26% for the total graduate population. Similar degrees of underrepresentation in the sciences exist for the Spanish-surnamed and American Indians.

Clearly, any program to achieve ethnic parity in scientific and technical fields must bring about a severalfold increase in minority graduate-student enrollment. This increase can be accomplished through a multiplicity of strategies and programs. Because graduate education is diverse, national in scope, and dispersed in terms of location of institutions and sources of students, it is helpful to divide the problem into several segments and to attempt to identify the principle factors that can be adjusted to raise the enrollment of minorities in the sciences. For convenience, three segments will be discussed: graduate schools, undergraduate schools, and secondary schools (junior and senior high schools). Each of these segments has a distinctive and equally important influence and role in determining the degree of minority participation in the sciences and engineering. The modification of one segment without complementary adjustments in the others would be inefficient and perhaps unproductive and futile.

Table 2. Representation of Minority Students in Each Graduate Field: All Institutional Respondents ($n = 154$)[a]

Field of study[b]	Total enrollment in each graduate field		Percentage minority in each field				
	Number	Percentage	Black	Spanish-surnamed	American Indian	Asian-American	Minority subtotals
Arts and humanities	53,920	100.0	2.8	1.5	0.3	0.9	5.5
Education	96,568	100.0	7.2	1.2	0.4	0.6	9.4
Engineering	31,273	100.0	1.2	0.8	0.1	3.3	5.4
Health professions	13,238	100.0	5.5	1.2	0.6	2.0	9.3
Life sciences	27,641	100.0	1.5	0.9	0.2	1.9	4.5
Biology	(5,027)	100.0	(2.6)	(0.7)	(0.1)	(1.7)	(5.1)
Biochemistry	(1,804)	100.0	(1.2)	(0.6)	(0.3)	(3.2)	(5.3)
Microbiology	(1,801)	100.0	(1.8)	(0.9)	(0.3)	(3.2)	(6.2)
Physiology	(1,110)	100.0	(1.5)	(0.9)	(0.3)	(2.0)	(4.7)
Other	(15,504)	100.0	(1.2)	(0.9)	(0.2)	(1.6)	(3.9)
Mathematical sciences	12,446	100.0	2.5	0.6	0.2	2.1	5.4
Physical sciences	21,629	100.0	1.4	0.7	0.2	2.6	4.9
Chemistry	(8,040)	100.0	(1.6)	(0.7)	(0.2)	(3.2)	(5.7)
Physics	(5,559)	100.0	(1.2)	(0.6)	(0.2)	(3.0)	(5.0)
Other	(6,560)	100.0	(1.2)	(0.7)	(0.2)	(1.5)	(3.6)
Basic social sciences	35,583	100.0	4.1	1.2	0.3	1.1	6.7
Economics	(5,766)	100.0	(1.9)	(0.8)	(0.3)	(1.6)	(4.6)
Psychology	(10,318)	100.0	(4.2)	(1.2)	(0.3)	(0.8)	(6.5)
Sociology	(4,566)	100.0	(5.8)	(2.0)	(0.2)	(1.3)	(9.3)
Other	(12,969)	100.0	(4.6)	(1.3)	(0.4)	(1.0)	(7.3)
All other fields	80,666	100.0	5.1	1.0	0.3	1.2	7.6
Total, all fields	372,964	100.0	4.4	1.1	0.3	1.4	7.2
Total U.S. population, 1970			11.1	4.4	0.4	0.4	17

[a]Based on data from the 154 Ph.D.-granting institutions able to provide minority enrollment data within field of study.
[b]Figures for subfields (in parentheses) sum to less than their respective field totals because some institutions reported data for the total field category but not for subfields.

Although it is recognized that a large number of additional areas could be identified as having influences on minority participation in the sciences—for example, elementary schools and family attitudes—they are not specifically addressed here in the belief that the three identified segments are the most malleable of the major determinants. Consequently, attention devoted to these major determinants will yield the greatest results in the shortest amount of time.

Graduate education in the sciences and engineering is highly dependent on funding—funding both for graduate students and for support of their research activities. In recent years, the federal government has reduced the levels of support of graduate education and research. This policy of reduced support is based on a number of considerations and includes the excessive production of Ph.D.'s as a central concern. Although the federal policy of reduced graduate-fellowship support and the elimination of training grants may be sound policy in the aggregate, it is not applicable to the situation of minority graduate-student enrollment and the production of minority doctorates. There is certainly no surplus of minority doctorates either in the sciences and engineering or in the humanities and social sciences.

Given the severe underrepresentation of minorities in the sciences and the need to expand the levels of production of minority doctorates, it is critical that the federal government, as the principal supporter of graduate education in the sciences, adopt a policy of directly encouraging and financially supporting the achievement of equal production of minority doctorates in the sciences and engineering. Without such a federal policy, the goal is probably unachievable by the turn of the century.

Two types of approaches can be employed. The first is a direct approach that will initiate programs of graduate fellowships and/or traineeships for the financial support of minority graduate students in the sciences, engineering, and economics and the other fields in which minority underrepresentation is exceptionally high. To be maximally effective, any program of support for minority graduate education must recognize that the centers of minority undergraduate population do not fully coincide with the major centers of graduate education. Graduate-student support programs must, therefore, include provisions for covering reasonable relocation costs and the out-of-state fees of state universities.

A second approach to the support of minority graduate education would be to give special recognition in the awarding of grants and contracts to those institutions with the best records of bringing minorities into the scientific professions and the best prospects for doing so in the future. For this approach to be most effective, it must be applied on a departmental as well as an institutional basis, as is implied by the experiences with the withholding of federal funds because of noncompliance with affirmative-action policies.

Table 3 provides a listing of universities in ranked order on the basis of their enrollment of black graduate students in 1972. Although these schools vary considerably in the scope and the characteristics of their graduate programs, they have one significant feature in common that distinguishes them from other universities. They all have external relationships and infrastructures that enable them to attract and retain

Table 3. 1972 Enrollment of Minorities in U.S. Graduate Schools[a]

	Percentages					
Institution	American Indian	Black	Oriental	Spanish-surnamed	Total enrollment	Federal support rank[b]
Howard University	—	73.3	—	—	1,148	8
Atlanta University	—	92.0	2.0		840	>100
Columbia University	0.5	6.0	2.7	1.8	4,972	12
University of Michigan	0.2	6.3	0.9	0.8	6,336	10
UCLA	0.5	5.6	5.2	4.4	6,232	2
Georgia State University	—	11.8	0.8	—	2,952	>100
Wayne State University	0.3	10.7	0.4	0.4	3,084	82
Harvard University	0.3	5.4	0.7	1.4	5,924	4
N. Carolina A. & T. State University	—	82.4	—	—	391	100
Rutgers University	—	10.9	2.4	0.6	2,606	64
U. C. Berkeley	0.2	4.0	3.6	1.6	7,073	11
Michigan State University	—	6.5	1.5	1.0	4,085	26
New York University	0.1	5.9	1.1	1.8	4,286	20
University of Chicago	—	6.3	1.2	0.9	3,900	16
Federal City College	—	67.8	2.2	0.2	351	>100
Stanford University	0.5	5.6	2.5	3.5	3,973	9
University of North Carolina, Charlotte	0.4	50.8	0.4	—	421	>100
University of Illinois	—	4.1	0.9	0.5	5,156	19

[a]*The Chronicle of Higher Education* (1974). Schools ranked by number of black graduate students.
[b]Ranking by level of federal support in fiscal year 1974. *Higher Education and National Affairs* (1975–1976).

200 or more black and other minority graduate students. These institutions, relative to other universities, are the most likely and the best able to accommodate increased numbers of minority graduate students. The awarding of grants, contracts, and traineeships to these institutions is likely to have a greater beneficial impact on minority graduate education than awards made to other institutions.

In a period of limited funding for graduate education and research, funding criteria must necessarily become more stringent. The evaluation criteria which must consider the potential social benefits accrued from the funding of one grant request as opposed to another could be included in a manner consistent with national policy without a significant distortion of the policy of awarding grants and contracts on the basis of merit and past performance. Of the 18 universities listed in Table 3, 10 are in the top 20 largest university recipients of federal funds (the extreme right column of Table 2 gives the federal-support rank). Obviously, as a result of inclusion of the consideration of minority enrollments in the criteria for the allocation of research grants and contracts, some institutions that are major recipients of federal funds would initially receive less support and others would receive more. However, it is also clear from the ranking of the universities listed in Table 3 that such a shift could occur without a significant, if any, reduction in the quality of work performed on grants and contracts.

At the collegiate level, a large number of areas can be identified for actions that would enlarge the pool of potential minority graduate students in the sciences. However, particular attention should be directed at retaining as science majors those minority students who enter college as declared science majors. Table 4 indicates that after four years of college, over 80% of the male students with science or engineering majors entered as freshmen with those declared majors. Thus, the most productive routes for increasing the output of minorities with bachelor degrees in the sciences and engineering should be based on increasing the numbers of minority students who enter as freshmen with declared science majors and increasing their retention rates in these majors.

Because scientific studies are more rigorous, and since science is seen to be more distant from our daily lives and cultural experiences than the humanities and the social sciences, a student, to be successful as a science major, must become acculturated in a "distinctive" subculture—a subculture that is rigidly based on logical and quantitative

Table 4. Shifts in Career Plans (1966–1970) of 1966 Male College Freshmen (in Percentages; $N = 462{,}239$)[a]

	Career plans in 1970[b]															
Career plans in 1966	1	2	3	4	5	6	7	8	9	10	11	12	13	14	15	16
1. Physician	39	6	4	—	3	—	5	1	7	1	3	10	2	—	14	6
2. Dentist	7	27	1	0	—	0	4	—	12	1	6	17	4	—	15	6
3. Research scientist	3	1	25	0	2	—	10	7	2	1	6	11	3	—	20	10
4. Statistician	0	0	3	0	0	0	4	0	4	3	5	53	0	—	26	4
5. Clinical psychologist	0	0	1	0	12	10	12	0	5	1	11	16	2	—	8	24
6. Social worker	0	0	0	0	2	24	6	—	4	1	7	16	2	—	32	6
7. College teacher	1	0	1	0	1	—	31	1	6	0	19	5	6	—	20	10
8. Engineer	1	—	3	—	—	—	3	41	3	1	5	19	2	—	15	7
9. Lawyer	1	0	—	0	—	1	6	1	47	0	3	16	2	—	19	5
10. Health professional	7	0	3	0	0	0	7	11	1	15	6	19	—	—	22	9
11. Elementary/ secondary-school teacher	—	—	1	0	—	2	11	—	1	1	41	10	3	—	23	7
12. Business executive or owner	1	—	1	0	1	1	5	2	8	—	3	45	3	—	23	7
13. Artist	—	0	1	—	2	—	8	4	2	1	9	11	39	—	13	10
14. Housewife	—	—	—	—	—	—	—	—	—	—	—	—	—	—	—	—
15. All others	1	1	2	—	—	1	5	3	5	2	6	18	3	—	44	9
16. Undecided	3	2	2	0	4	1	5	6	6	—	10	12	4	—	23	21

[a]Information is from Padulo (1974).
[b]Numbers under "Career plans in 1970" refer to the same categories listed under "Career plans in 1966."

analysis, the precise use of language and terms, and long, tedious hours of problem sets and laboratory work. For minority students, an undergraduate science major—particularly one not directed toward medical or dental school—may not appear to be sufficiently rewarding.

Majoring in a science may be particularly unfeasible for students with major financial concerns. Table 5 indicates that minority students are generally at a greater financial disadvantage than the average college student. In 1971, over 70% of minority freshmen came from families with incomes of less than $10,000, while only 34% of all freshmen came from such families. Clearly, financial considerations must have a major influence on minority students' enrollment in college and on their choice of majors. Financial considerations lead to the selection of those majors that represent the greatest financial security, especially to low-income students. In the technical fields, only engineering offers a relatively quick route to financial security—a bachelor's degree in engineering is a professional degree that ensures a relatively high starting salary. In 1974, monthly starting salaries for recent graduates with bachelor's degrees were: engineers, $964; sciences and health professions, $821; and humanities, social sciences, and business, $786.

To address the twin concerns of the financial needs of minority students and their retention in science, funds to support undergraduate science majors working in laboratories either on work–study programs or undergraduate research projects should be expanded substantially—particularly in those institutions with large enrollments of minority undergraduates. Funds that support these students working in labora-

Table 5. Family Income Distribution in the United States, 1971

	Percentage			
Income	Population as a whole[a]	Blacks[a]	Freshmen in all colleges[b]	Minority freshmen in all colleges[c]
Below $6,000	24.2	46.9	12.0	43.9
$6,000–$9,999	24.0	25.4	22.4	27.7
$10,000–$14,999	26.9	17.2	32.3	17.3
$15,000 and over	24.8	10.6	33.3	11.0

[a] Source: U.S. Department of Commerce, Bureau of the Census (1972, Table 16, pp. 40–41).
[b] Source: *The American Freshman* (1971)
[c] Source: Bayer (1972, p. 39). The data are from 1971. Figures are national norms for blacks only, but a survey by Bayer (p. 85) of Chicanos, Puerto Ricans, and American Indians indicates that the patterns for the other underrepresented minorities are similar.

tories have the obvious double benefits of addressing the financial needs of the students and encouraging their immersion in science and scientific research. The Minority Biomedical Support Program of the National Institutes of Health, which provides support for undergraduate participation in research in schools with large minority enrollments, is an excellent model for this type of program. Similar programs should be established for the physical sciences and engineering.

In a detailed analysis of *Minorities in Engineering: A Blueprint for Action* (Padulo, 1974), it is proposed that as part of a comprehensive effort to achieve parity of minority enrollments in freshmen engineering classes by 1982 and in all classes by 1987, the six traditional black colleges (Prairie View, Howard University, Southern University, North Carolina Agricultural and Technical College, Tennessee State, and Tuskegee), which have produced over half of all black engineers, double their enrollments. If a similar detailed analysis is made for the basic sciences, it would probably produce a similar approach, that is, the expansion of science programs and enrollments in the traditional black colleges, concurrent with expanded minority enrollment in all undergraduate schools.

The junior- and senior-high-school years represent the critical formative years for potential scientists in terms of both career interest and academic preparation. Table 6 indicates that the preponderance of freshmen science and engineering majors had decided on scientific or technical careers by the ninth grade. The combination of this information with that of Table 4 indicates that the majority of scientists with bachelors' degrees had chosen a scientific or technical career by the ninth grade, leading to the conclusion that efforts to stimulate an interest in science must be directed at minorities prior to the ninth grade. Since the junior-high-school years represent an age period in which students grow substantially in their maturity, understanding of reality, and degree of self-direction, this age group would appear to be the most receptive to information on science and engineering careers and advice on how to prepare for such careers. The junior-high-school years also represent a three-year block of time that can be used to bring the basic academic skills (reading, writing, and arithmetic) of students up to levels sufficient for the pursuit of a high-school academic program in the sciences.

For all undergraduate engineering majors, the principal factors leading to attrition were academic problems, 45%; discord with profes-

Table 6. Proportion of High-School Boys Shifting from Each Field between the Ninth Grade and One Year after High School[a]

Field choice in the ninth grade	Field choice one year after high school										
	1	2	3	4	5	6	7	8	9	10	11
	(Percentages)										
1. Natural sciences	12.4	10.7	9.7	6.4	3.4	1.0	3.4	10.4	15.2	3.8	23.5
2. Engineering	5.0	20.0	13.0	3.0	3.0	1.0	2.0	8.0	21.0	1.0	23.0
3. Business	3.7	3.6	35.1	1.9	2.6	1.2	2.3	8.1	16.0	1.4	24.0
4. Health—M.D. and D.D.S.	2.4	5.6	12.7	26.3	5.4	0.6	2.2	8.3	14.4	3.2	18.8
5. Law	2.0	4.0	16.0	4.0	20.9	2.0	3.0	11.0	13.0	2.0	21.9
6. Clergy	3.1	3.1	8.9	3.1	5.1	28.8	3.1	11.0	12.0	4.1	17.9
7. Arts	1.6	1.8	10.2	1.3	1.8	1.0	25.4	11.5	13.6	3.7	28.0
8. Teaching	3.0	3.0	13.0	3.0	2.0	1.0	3.0	27.0	17.0	4.1	23.9
9. Skilled/technical	3.1	5.8	12.6	1.5	1.7	0.6	2.0	7.0	34.0	1.5	30.2
10. Social sciences and services	3.6	1.8	13.8	1.8	5.4	1.8	3.0	6.0	20.3	3.0	39.5
11. Other	4.0	6.0	14.0	2.0	2.0	1.0	5.0	9.0	23.0	1.0	33.0

[a]Source: Folger, Astin, and Bayer (1970, Table 6.2).

sors or courses, 40%; personal and family problems, 10%; and financial problems, 5% (Padulo, 1974, p. 81).

The minor importance of financial problems to the attrition of all students probably does not hold for minority students, who are twice as likely to come from low-income families. The areas of most frequently reported academic weaknesses and high-school deficiencies were mathematics, 47%; physical sciences, 33%; natural sciences, 12%; humanities, 5%; and social sciences, 3% (Padulo, 1974, p. 82).

In the absence of more specific data, it is reasonable to assume that similar characteristics hold for the sciences in general and for minority students in particular.

The relatively lower rates of minority-student enrollment in technical fields is probably a reflection of their poor level of preparation in mathematics and science. Basic skills and the development of interests in technical careers in mathematics and science are established in junior and senior high schools. The basic mathematics curriculum in secondary schools follows the general pattern outlined in Table 7. The fundamental importance of algebra and geometry is indicated by the several tracks that students can take to complete these subjects at any time from the 9th through the 12th grades and the percentage of their contributions to the questions on the mathematics sections of the college-placement examinations. Unfortunately, minority students in many school districts do not complete these basic subjects at the same rates as other students. Nor do they complete, at sufficient rates, the third year of mathematics—advanced algebra and trigonometry—recommended as preparation for freshmen calculus and science majors.

Table 8 provides data on the lower rates of enrollment in mathematics courses of minority students in a single California school district. Since algebra is a prerequisite for college-preparatory chemistry and geometry is a prerequisite for college-preparatory physics, efforts to increase minority-student enrollments in these two basic mathematics courses (as well as a third year of mathematics) must be central to any program designed to achieve increased minority participation in the scientific and technical fields. Without vigorous national and local programs to increase the enrollments of minority secondary-school students in algebra and geometry (as well as a third year of high-school math), programs at the college and graduate levels will be restricted in their effectiveness because of a limited pool of prepared minority students.

Table 7. Patterns for Secondary Mathematics

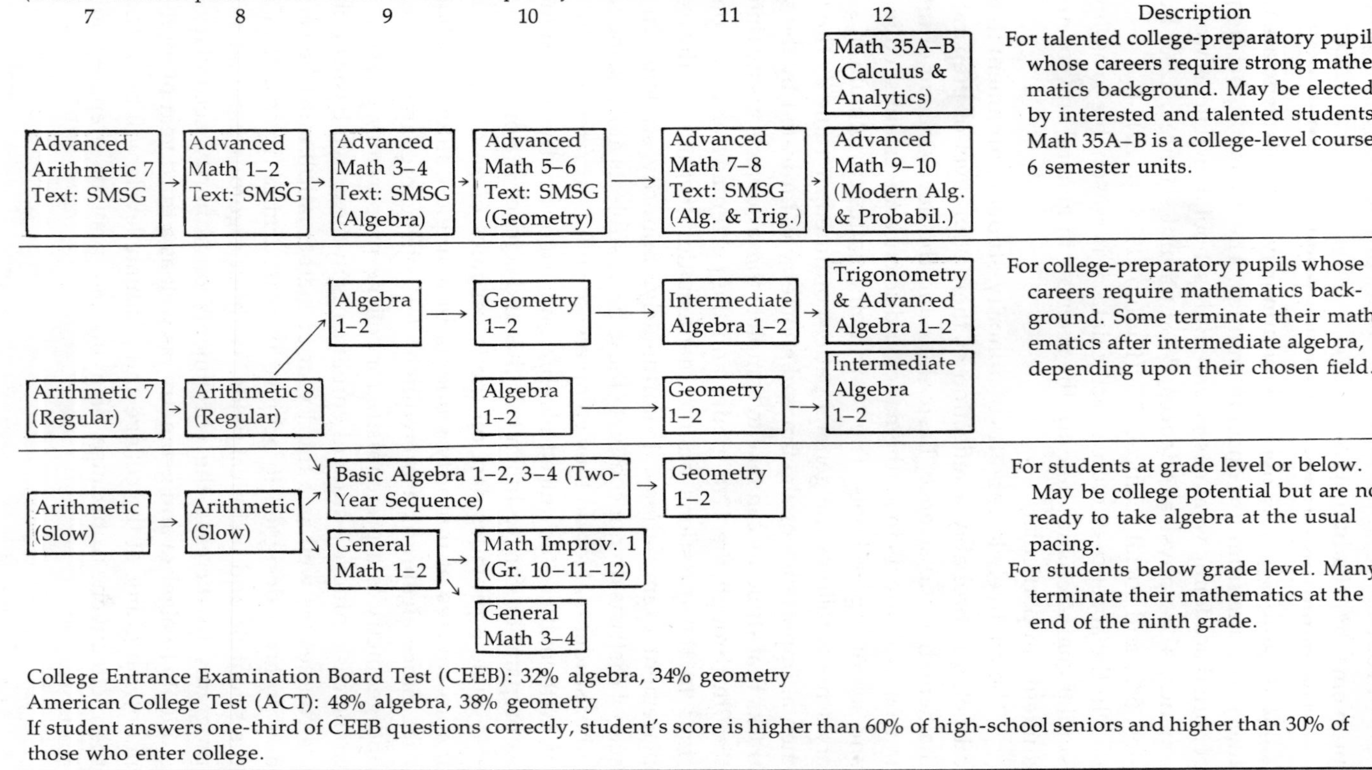

(Arrows indicate places where students most frequently move to other levels)

7	8	9	10	11	12	Description
					Math 35A–B (Calculus & Analytics)	For talented college-preparatory pupils whose careers require strong mathematics background. May be elected by interested and talented students. Math 35A–B is a college-level course, 6 semester units.
Advanced Arithmetic 7 Text: SMSG →	Advanced Math 1–2 Text: SMSG →	Advanced Math 3–4 Text: SMSG (Algebra) →	Advanced Math 5–6 Text: SMSG (Geometry) →	Advanced Math 7–8 Text: SMSG (Alg. & Trig.) →	Advanced Math 9–10 (Modern Alg. & Probabil.)	
		Algebra 1–2 →	Geometry 1–2 →	Intermediate Algebra 1–2 →	Trigonometry & Advanced Algebra 1–2	For college-preparatory pupils whose careers require mathematics background. Some terminate their mathematics after intermediate algebra, depending upon their chosen field.
Arithmetic 7 (Regular) →	Arithmetic 8 (Regular) ↗ (to Algebra 1–2)		Algebra 1–2 →	Geometry 1–2 →	Intermediate Algebra 1–2	
	↘	Basic Algebra 1–2, 3–4 (Two-Year Sequence) →		Geometry 1–2		For students at grade level or below. May be college potential but are not ready to take algebra at the usual pacing.
Arithmetic (Slow) →	Arithmetic (Slow) ↗ ↘	General Math 1–2 → ↘	Math Improv. 1 (Gr. 10–11–12)			For students below grade level. Many terminate their mathematics at the end of the ninth grade.
			General Math 3–4			

College Entrance Examination Board Test (CEEB): 32% algebra, 34% geometry
American College Test (ACT): 48% algebra, 38% geometry
If student answers one-third of CEEB questions correctly, student's score is higher than 60% of high-school seniors and higher than 30% of those who enter college.

Table 8. Mathematics Data for Selected High Schools in a Single District[a]

High schools	Number of seniors per school	Percentage minority	Percentage in geometry	Percentage in advanced algebra and/or trigonometry
Average of 5 schools	1000	95	23	12
Average of 2 schools	900	32	42	32
Average of 3 schools	1100	5	49	42
Districtwide		49	40	21

[a]These data were collected by the author in cooperation with a large metropolitan school district.

The underrepresentation of minorities in the scientific and technical professions represents the wasting of an untapped intellectual resource that our country can ill afford. The achievement of greater minority participation in these professions should not be viewed in isolation but as part of a dedicated national commitment to make the American dream a reality for all during the last quarter of the twentieth century.

REFERENCES

The American freshman: National norms for Fall 1971. Washington, D.C.: American Council on Education, 1971.

Bayer, A. E. *The black college freshman.* Washington, D.C.: American Council on Education, 1972.

El-Khawas, E. H., and Kinzer, J. L. *Enrollment of minority graduate students at Ph.D. granting institutions.* Higher Education Panel Report, No. 19. Washington, D.C.: American Council on Education, 1974.

Folger, J. K., Astin, A. W., and Bayer, A. E. *Human Resources and Higher Education.* New York: Russell Sage Foundation, 1970.

Padulo, L. (Chairman). *Minorities in engineering: A blueprint for action.* A report of the Planning Commission for Expanding Minority Opportunities in Engineering. New York: The Alfred P. Sloan Foundation, 1974.

U.S. Department of Commerce, Bureau of the Census. *Money income in 1971 of families and persons in the United States,* Current Population Reports, Series P-60, No. 85. Washington, D.C., December 1972.

12

National Institute of Mental Health Research and Training Policies Affecting Minorities: An Outsider's View

JEAN V. CAREW

My purpose today is to present an outsider's view of the National Institute of Mental Health's research and training policies as they affect minorities. From the point of view of an outsider, the most significant change in the NIMH's policies affecting minorities has been the establishment of the Center for Minority Group Mental Health Programs. The center was founded in November 1970 to serve as a focal point for all activities within NIMH bearing directly on meeting the mental-health needs of minorities. The founding of the center was not a sudden act of benevolence. Rather, it was the culmination of two years of pressure and outspoken criticism of NIMH by the Black Psychiatrists of America for their failure to eliminate racism within the NIMH and to mount any significant effort to eradicate racism in America. The mental-health implication of racism had been recognized and documented in two major reports published the preceding year, the first by the Kerner Commission (National Advisory Commission on Civil Disorders, 1968) and the second by the Joint Commission on Mental Health of Children (1968). In the latter report, the Committee on Minority Group Children agreed: "Racism is the number one public health problem facing America today. The conscious and unconscious attitudes of superiority which permit and demand that a majority oppress a minority are a clear and present danger to the mental health of all children and their parents." This belated recognition of the direct relationship between mental health and racism owed much to the views

JEAN V. CAREW • Graduate School of Education, Harvard University, Cambridge, Massachusetts.

and previous writings of the founders of the Black Psychiatrists of America (Pierce, 1969, 1973, 1974). It is important to realize that the group came into being only in 1969. It took enormous courage, sagacity, and commitment on the part of its leaders—notably Dr. Chester Pierce of Harvard University and Dr. James Comer of Yale—for this group to confront the NIMH bureaucracy directly. Their demands were forceful and specific. First, they wanted a commitment from the NIMH to establish an identifiable unit within the NIMH bureaucracy to promote research and training programs in mental health for minority groups, and second, they requested that the NIMH make a significant effort to reduce racism within its own institution and grants. Eighteen months later, the first of these requests was implemented in the founding of the Center for Minority Group Mental Health Programs at the NIMH.

In its program development, the operational frame of reference of the center has been the definition of racism stated in the U.S. Commission on Civil Rights (1970) report, *Racism in America and How to Combat It*. The report states: "Racism may be viewed as any attitude, action or institutional structure which subordinates a person or group because of his or her color. . . . [Racism] is not just a matter of attitudes: actions and institutional structures, especially, can also be a form of racism."

As seen by the center, the problem is clearly *white racism,* and in particular, white-controlled *institutional racism.* Its solution requires radical, pervasive, and long-term change on at least three fronts: (1) developing knowledge of minority-group cultures with special emphasis on their strengths, adaptability, and coping skills; (2) developing knowledge of white racism and of strategies to eliminate it in official institutions; and (3) increasing minority-group mental-health and behavioral-sciences manpower, including research and service capabilities. These objectives are addressed directly by two programs of the minority center: the Minority Group Program and the Racism and Mental Health Program.

RESEARCH PROGRAMS

The center seeks to support research on the positive aspects of minority-group mental health and research that challenges traditional assessment and treatment techniques. One strategy used by the center toward this end is the support of national research-and-development centers for four minority groups: one for American Indians, two for black Americans, one for Asian-Americans, and one for Spanish-speaking Americans. Plans for a second Spanish-speaking center are underway. These centers have three main components: scholars in residence, technical assistance, and data banks. They have all been fairly recently established and are in the initial stages of conceptualizing and generating research.

The center has also funded a number of research studies by individual investigators on topics of priority concern. For example, the objective of one study now being supported is to construct a reliable and valid intelligence test for black children. Another project examines predisposing and contributory factors in Asian-American and black American mental-health problems with the aim of designing more effective treatment models for these groups. A third study investigates the relationship of suicidal behavior in four American Indian cultures to the stress of urbanization and acculturation. Equally important in the center's priorities is research on institutional racism and strategies to eradicate it. Prototype studies on racism in higher education, in law enforcement, and in state mental-health institutions are now being funded or are under review.

About 40% of the center's budget is devoted to increasing minority manpower in mental-health research and service through training grants. At present, grants have been made to support graduate fellowships for minority students through five professional associations (the American Psychiatric Association, the American Psychological Association, the American Sociological Association, the Council on Social Work Education, and the American Nurses Association).

The center has also been very active in seeking guidance and feedback from the field to identify needs, establish objectives, and evaluate programs. Initially, conferences were held for each of the four minority groups, and from these have emerged coalitions (e.g., the Pacific Asian Coalition and the Coalition of Spanish-Speaking Mental Health Organizations) whose purpose is to organize and articulate the needs of minority constituencies. In-house minority influence on center policy and allocation of resources is assured by an initial-review committee composed predominantly of minority individuals.

So far, we have focused on one small unit of the NIMH: the minority center. The center has great symbolic importance, and given the constraints of its small budget and staff, it has achieved remarkable accomplishments. But the very size of its funding—less than 4% of total NIMH expenditures—is sufficient to tell us that its mission cannot be realized by its efforts alone. It is crucial therefore to examine the NIMH's overall policies and practice on the research and training of minorities and the role played by leaders in the minority center in bringing about change.

As far back as 1971, Dr. Bertram Brown, on assuming the directorship of the NIMH, set minority mental health as one of the three top priorities his administration would pursue. Has this commitment gone beyond rhetoric? Let us examine the evidence.

The NIMH, as we all know, is *the* federal agency administering mental-health programs. Its basic mission is to develop knowledge, manpower, and service to treat and prevent mental illness and to promote mental health. Its impact can hardly be overstated, since directly or indirectly the NIMH probably affects the delivery of all mental-health services in the United States. With the likely establishment of a national health-insurance program and the implementation of a national network of community mental-health centers within the next few years, the NIMH's future role in the mental health of the public will become even larger. These two predictable changes imply an escalating need for minority mental-health knowledge and manpower. Minorities represent more than 17% of the population, and their oppressed condition is associated with disproportionately high levels of alcoholism, drug abuse, mental-health problems, and institutionalization in mental-health facilities. Yet, minority mental-health power continues to be shockingly low, and there is abundant evidence that treatment models and delivery systems fall far short of meeting minority needs (see, e.g., Sue, Allen, McKinney, and Hall, 1974; Yamamoto, James, and Palley, 1968).

The most powerful role played by the NIMH vis-à-vis minority mental health is at the level of *resource allocation*. The NIMH exerts its impact on mental health not so much by the research and training it carries out in-house but much more through the system of grants and contracts that it awards to institutions and individuals outside the agency (e.g., universities and hospitals). Thus, so far as minority health is concerned, questions of who receives NIMH resources (e.g., the proportion of minorities in application-review groups and support staff) and what criteria are used in the rating of applications (e.g., appropriateness of research or training curriculum to minorities and proportions of minority subjects, trainees or clients) are all of critical importance.

The interdependence of these aspects of NIMH policy and practice regarding minorities has long been stressed by key minority administrators within the NIMH, such as the director of the minority center, Dr. James Ralph, and the present director of the Division of Manpower and Training Programs, Dr. William Denham. Both have made strenuous efforts to bring about change in each of these areas. Dr. Ralph, for example, has for years requested that each division collect and report information on its allocation of resources to minorities. Yet, as late as 1974, reliable information from most major divisions was still not forthcoming. Thus, in his report to the National Advisory Mental Health Council in 1974, Ralph noted that the major division responsible for allocating research grants, the Division of Extramural Programs, was unable to supply information about the proportion of research funds going to minority investigators. It simply had not collected this information in any systematic fashion, although it did not deny that the percentage was probably very small. On training grants, the major division, Manpower and Training Programs, was more helpful. Their data showed a steady increase in grants to minority training directors from 4.7% in 1971 to 8.6% in 1974, but the performance of various sections within the division was exceedingly variable. Some sections, such as New Careers, Continuing Education, and Social Work, awarded minorities a fairly equitable share of their resources; others with large budgets, like Psychiatry and Psychology, had not (less than 3% of their funds went to minorities). Similar patterns were found for research and training grants categorized in terms of degree of minority concern relevance rather than in terms of the race of the immediate recipient. These data must be accepted cautiously, since a research grant may be "concerned with" minorities yet detrimental to them, and similarly a training grant may involve many minority trainees yet train them inappropriately. Even so, the data indicated that that funding of research grants centrally concerned with minorities by the Division of Extramural Research used no minority consultants, while over the same period about 25% of the consultants engaged in Manpower and Training were members of minorities.

What now of the people who make the decisions allocating NIMH resources, the members of initial-review committees and the NIMH review staff? The statistics are a bit more encouraging. The percentage of minority-group members on the initial-review committees advising the Division of Extramural Research was 7.6% in 1971 and increased to 11.6% in 1973. The corresponding percentages in Manpower and Training were 29.7% in 1971 and 25.1% in 1973. Regrettably, data on the minority composition of staff who play key roles in the decision-making process were not available.

These figures point, if nothing else, to the urgent need for the NIMH to create an efficient data-collection system regarding its performance vis-à-vis minorities. One way in which institutions perpetuate racism is by evading or deliberately failing to collect information on their own practices. Thus, Ralph's report to the advisory council failed to receive clearance in written form because of alleged errors, inconsistencies, and incompleteness of data, even though the responsibility for those inadequacies clearly lay with the reporting divisions. It has taken over one year—and many negotiations between Ralph and the NIMH administration plus another report, this time by an outside consulting firm at considerable cost (Finan, Nordlie, Witten, and Sinnot, 1975)—for a comprehensive system of collecting such data to be seriously considered. The adoption of such a system is incorporated in the latest action plan (January 1976) formulated by the director of the NIMH, which is outlined below.

But documenting unequal opportunity is one thing, rectifying it is another. On the basis of the findings presented in his 1974 report to the council, Ralph proposed that the NIMH take immediate action to implement its minority priority by targeting a minimum of 20% of research and training funds in fiscal year 1976 for minority projects. This was hardly a radical proposal, since a figure of 25% had often been cited by NIMH director Bertram Brown as an equitable proportion of resource allocation to minorities. Furthermore, a precedent for the targeting of research monies for a specific priority had already been established under his administration in the case of child mental health. Nevertheless, Ralph's proposal was delivered to the NIMH at a time of rapidly declining budget, impounding of funds, and general fiscal gloom, and it aroused intense debate. Out of this debate, the action plan previously referred to finally emerged. In this plan, the director of the NIMH proposed a three-phase, three-year effort to increase minority-related research and research training. The essentials of his plan follow.

In Phase I (to start immediately), the directors of each research division are instructed to submit (1) a description of the division's research priorities regarding minority-focused research and (2) an indication of the amount and percentage of turnover funds (new money) targeted for such research. These amounts are to be regarded as targets for planning rather than as a hard-and-fast earmarking of funds, and the achievement of target goals is seen to depend on the influx and the quality of grant applications. Minority projects will undergo the same review process as other applications, and only those falling within the normal range of acceptable review will be eligible for funding.

In Phase II, concurrent with the development of 1976 division plans, a series of workshops and conferences are to be held to further identify gaps in knowledge regarding minority mental health, to formulate research efforts, and to develop appropriate methodology. Concomitantly, technical assistance to minority researchers is to be expanded to ensure high-quality research applications.

Finally, Phase III calls for the establishment of a monitoring system by March 1976 to assess the current status of resource allocation to minorities and the achievement of minority target goals. In this action plan, the NIMH seems at long last to be taking genuine steps to implement its minority-research priority, which its director embraced five years ago. But will this NIMH "action plan" be acted on? It is hard to be hopeful, considering the current gloomy picture with respect to funding and, in particular, the fiscal year 1977 budget, which proposed a phaseout of all NIMH support for clinical training and a draconian reduction of support for research training. In such a climate, all our past experience tells us that it will be the minority priority that will be the first to go unless strong men and women both inside and outside the NIMH have the will and the courage to fight.

REFERENCES

Finan, B., Nordlie, P., Witten, D., and Sinnot, J. Development of quantitative indices of institutional change with regard to racial minorities and women in NIMH external programs. Prepared under contract to NIMH, 1975.

Joint Commission on Mental Health of Children. Report. Washington, D.C.: U.S. Government Printing Office, 1968.

National Advisory Commission on Civil Disorders. Report, Otto Kerner (Chairman). Washington, D.C.: U.S. Government Printing Office, 1968.

Pierce, C. M. Is bigotry the basis of the medical problems in the ghetto? In J. C. Norman (Ed.), *Medicine in the ghetto*. New York: Appleton-Century-Crofts, 1969. Pp. 301–312.

Pierce, C. M. The formation of the black psychiatrists of America. In C. Willie, B. Brown, and B. Kramer (Eds.), *Racism and mental health.* Pittsburgh: University of Pittsburgh Press, 1973. Pp. 525–553.

Pierce, C. M. Psychiatric problems of the black minority. In S. Aneti and G. Caplan (Eds.), *American handbook of psychiatry,* 2nd ed. New York: Basic Books, 1974. Pp. 512–523.

Sue, S., Allen, D., McKinney, H., and Hall, J. *Delivery of community health services to black and white clients.* Seattle: University of Washington Press, 1974.

U.S. Commission on Civil Rights. Racism in America and how to combat it. Washington, D.C.: U.S. Government Printing Office, 1970.

Yamamoto, J., James, Q., and Palley, N. Cultural problems in psychiatric therapy. *Archives of General Psychiatry,* 1968, *19,* 45–49.

13

An Informed Constituency with a Representative Bureaucracy: Health Policy and Black People

THERMAN E. EVANS

It has been said that "Those who define are masters." I believe this is true since inherent in the role of "definition maker" is the authority and power to make definitions; and if one is in a position to determine what is, certainly he or she is in a better position than most to determine what will be. There are other implications inherent in this statement. There are implications for those who are defining and for whatever or whoever is being defined. One thing is certain, if you are in a position to make definitions, if you can determine the results of anything, usually those results will be reflective of who and what you are.

What does this mean with respect to health policy and black people or to health policy and Spanish-speaking or American Indian people? It simply means that we have not been in positions of power or policy-making positions, and as a result, we have not realized our "fair share" or even a representative share of the health-related programs. This is indicated by the priority given to programs, which in turn determines the flow of money into the medical-care system. For example, if this is the year for hyaline-membrane disease or emphysema, this will be reflected by the concentration of dollars in research, consumer-health information, and medical-care delivery for hyaline-membrane disease and emphysema. The areas of need that receive dollars, and how much they receive, are all determined by the policy makers, for they define and they determine the outcome.

Another case in point is the enrollment of black students in this

THERMAN E. EVANS • Executive Secretary, Health Manpower Development Corporation, Washington, D.C.

country's medical schools. As of school year 1974–1975, there were 3,396 (6.3%) black students enrolled out of a total of 54,074. Placed in the context of the small number of black physicians—approximately 6,000 (1.8%) out of a total of approximately 360,000—the black population in medical schools does not move us very far toward an equitable representation of black physicians. It should be pointed out that approximately one-fourth of the 3,396 black medical students enrolled in 114 medical schools were enrolled in Howard University and Meharry Medical College, schools that have been responsible for training at least 75% of all currently practicing black doctors.

Why is this so? Why is it that two schools can accept and matriculate 25% of all the black medical students and the other 75% are spread among 112 institutions that are, in most instances, larger and better funded? The answer simply is that there have not been enough black people or sensitive whites in the positions of power, in the definition-making roles, or on the policy boards that determine outcomes (i.e., who gets admitted) at these institutions.

Obviously, since there are so few black health professionals, there will be only a limited number from which to select to serve in any policy-making body. The result is minimal representation of blacks at health policy-making levels, which has led to past and current neglect of the many health problems that poor people in general and black people in particular face in this country.

The effect of the inadequate representation of black people at the health policy-making levels is further detailed in a 1971 memorandum on "The Health Status of America's Non-Whites." This memo was prepared for the Undersecretary of the Department of Health, Education, and Welfare for transmission to the Secretary of HEW, who at that time was Elliot Richardson. The memo said:

> It has been recognized that a contributing cause to the relative poor health of non-whites is the inadequate numbers and maldistribution of health manpower and the nonavailability of, or poor accessibility to, health service facilities. . . . However there is an additional probable cause which deserves full consideration and should be of direct interest, namely, *an insufficient non-white perspective in decision-making on health matters*. . . . Historically, in this country minorities, not being in positions of influence have had to ultimately rely on the goodwill of the white majority, hoping that it would act morally. This condition has applied to health matters as well. The question at the moment is whether minorities can or should continue this reliance.

History has shown that the white majority has not acted morally; indeed, in many instances when the health of black people was the issue, the white majority has not acted at all. Present data on the health status of blacks and on the number of blacks participating in setting health policy make it clear that we should learn from history. For example, the same 1971 memorandum on "The Health Status of America's Non-Whites" can be referred to again. The memo quoted a passage that accurately assessed the general health status of black people today:

> The Committee's survey does not include data for Negroes. It is well known, however, that the 10% of our population who are colored have health problems which are, on the whole, considerably more serious than those of whites. The Negro is America's principal marginal worker, and he suffers in the North as well as the South from the many disabilities that this entails: poor housing, less adequate diet, less sanitary surroundings, more employment of married women, and greater economic insecurity.
>
> The extensive migrations of Negroes during the last 20 years have added new complications to their problems. Although Negroes have lower death rates than whites for a few diseases, rates double the rates of whites are recorded for tuberculosis, organic heart disease, acute and chronic nephritis, cerebral hemorrhage, pneumonia, typhoid fever, whooping cough, bronchitis, puerperal conditions, influenza, malaria, and pellagra. Not only are death rates higher but so is the incidence of illness, at least from certain diseases. Syphilis, for example, is nearly two and a half times as prevalent among Negroes as among whites. Because Negroes generally are poorer than whites, it is safe to assume that on the average, they receive less medical services.

This quotation was extracted from a report by the Committee on the Costs of Medical Care, which did its report on medical care in this country in 1932, over four decades ago. The relative health status of blacks is unchanged.

Probably the most significant reason for the relatively unchanged health picture of the United States minorities in general, and black people more specifically, is the lack of the concentrated resources necessary to address their more prevalent and severe health problems. The required resources, both historically and currently, have not been available because health-policy makers, those who determine the priorities in health, characteristically have not come from the black population; nor have whites with a sensitivity to these problems been represented on these policy-making bodies in sufficient numbers to make significant changes.

As a further illustration of the effect of a lack of input from the black perspective, data from the National Institutes of Health (NIH) can be examined. According to the fiscal year 1974 annual report of the NIH, 17,327 research grants were awarded totaling $1,329,892,660. Of the total number of grants, the 112 predominantly black institutions received 99 (0.6%) (U.S. Department of Health, Education, and Welfare, 1975b). Of the total dollar amount, the 112 predominantly black institutions received approximately $10,400,000 (0.8%). The same report indicated that the University of Alabama, for example, received more money—$11,429,115—than did all of the 112 predominantly black institutions combined. Why are these black institutions receiving such a disproportionately small amount of tax dollars when these dollars are most crucial for the existence of all educational institutions?

This question forces us to look at the people in "definition-making" positions at the NIH in an attempt to determine patterns that may be responsible for the current picture. The advisory panels, review committees, or advisory councils are the consultant bodies that made the decisions on where money goes, how much, and to whom. These august bodies meet once a year, twice a year, and sometimes three or four times a year, or when there is a series of applications for review. As of mid-1975 (the latest assessment available at this writing; all figures are approximately as provided by the Equal Employment Opportunities office), the NIH advisory panel was composed as follows:

Total number of people serving:	1,884
Total number of U.S. nonwhites: (Includes Spanish-surnamed, American Indians, Asian-Americans, and blacks, who constituted 95% of this nonwhite group)	119

Clearly, these proportions depict an underrepresentation of minorities at the decision-making levels of the National Institutes of Health.

The 1974 employment data on the ethnic makeup of employees at GS 12–17 levels in the NIH (U.S. Department of Health, Education, and Welfare, 1975a) further illustrate the point (see Table 1). GS levels 12–

Table 1. National Institutes of Health, Ethnic Breakdown of Employees at GS 12–17 Levels as of 1974

Grade	Black			American Indian		Spanish-speaking		Asian-American		White		Total	
level	Male	Female	%	Male	Female	Male	Female	Male	Female	Male	Female	minority	Total
GS–12	32	16	10	0	0	3	2	9	2	273	178	64	515
GS–13	20	8	5	0	0	3	1	12	2	432	106	46	584
GS–14	10	3	3	0	0	4	1	15	1	389	72	34	495
GS–15	6	1	2.5	0	1	2	0	8	2	291	40	20	351
GS–16	2	0	2	0	0	0	0	2	0	80	2	4	86
GS–17	0	0	0	0	0	0	0	1	0	17	0	1	18

17 were selected since these represent positions that generate and implement internal and external policies. Needless to say, programs will reflect the interest of those at these hierarchical levels. Hence, the flow of dollars will also coincide with the same lines of interest.

The question can be asked, does the simple addition of professionals and nonprofessionals who happen to be black to policy boards that determine outcome produce a significant change in what the outcome will be? I feel the answer to this is obviously yes. The Institute of Medicine of the National Academy of Sciences (NAS) produced a 1974 report called "Changing Values and the Selection Process for Medical School" (NAS, 1974). This article contained many interesting points that corroborate one of my main themes; that is, if the makeup of the policy-making body is changed, then the policies that body makes will change, and thus the outcome will be different. In the NAS report, Wellington said, "Few would argue against the notion that a committee [admissions committee] chooses its own image; as such committee members are selected because they hold the given sets of values, these values will very likely be reflected in the choice of candidates." He went on to say, "The result of increased numbers of minority persons and women on the [admissions] committee was an increase, by nearly the same percentage, of minority and women students in the entering class." There is a specific example to illustrate further the importance of the nonwhite perspective in decision-making on health matters. In September 1970, Colbert King, a 33-year-old government employee and a native of Washington, D.C. was a HEW fellow working with Undersecretary John G. Veneman. Because of an inquiry directed to HEW, King was asked to explore what was being done about sickle-cell anemia. In response to that request, Mr. King:

1. Talked with doctors at Howard University.
2. Visited the NIH at Bethesda, Maryland, where he was told "that the NIH attempted to address the problems of greatest concern to the largest number of people."
3. Was assured that much basic research had been done on sickle-cell disease (which was true), although NIH research obligations for sickle-cell anemia never exceeded $250,000 until 1970. During 1967–1971, $11.8 million was allotted to cystic fibrosis, a disease mainly afflicting whites (who are 90% of the population) with about 1,200 new cases a year—only a few more than sickle-cell anemia (which mainly affects 10% of the population).

4. Was told that the molecular mechanism of sickle-cell disease was well understood, although the knowledge had not yet produced any cures.

In any case, Mr. King recommended a new program of medical training, public education, and research on sickle-cell anemia as well as "equitable" representation on the NIH advisory groups responsible for reviewing grant requests. Among 113 such advisers, he found 1 black.

Mr. King got what he referred to as "strong, active support" from Veneman's assistant Russell Beyers, Deputy Undersecretary Robert Patricelli, Undersecretary Veneman, and Secretary Richardson. Patricelli also inspired his father, Leonard J. Patricelli, President of WTIC-TV in Hartford, Connecticut, to produce a pioneering and influential series of television programs on the disease.

On February 12, 1971, the Assistant Secretary for Health, Dr. Robert O. Egeberg, ordered the NIH to "articulate" a "clearly visible" program for sickle-cell anemia within five days. On February 19, Mr. Nixon publicly promised a $6-million budget to make this an officially "targeted" disease for concentrated research and other efforts; Congress gave him $10 million.

Colbert King, who triggered the new multimillion-dollar campaign against sickle-cell anemia, has received almost no public credit. However, the experience of Mr. King is an exciting example of how one black man, a minute fragment of the minority population, can identify a problem, suggest a solution, and prod the bureaucracy toward implementation.

Clearly, this specific example speaks to what can happen when the ethnic perspective at the policy level is changed just a little bit. It also, along with the other information I've presented here, makes a good case for a more representative bureaucracy. I do not see the bureaucracy itself as a problem. There must be some mechanism for running the government, and a bureaucracy works very well for most groups who are represented by it. Herein lies the crux of the problem for Mexican-American, Puerto Rican, American Indian, and black American citizens. These groups are underrepresented at policy-making levels in the bureaucracy.

This underrepresentation, which borders on virtual exclusion, and a generally uninformed constituency have resulted in and continue to result in the neglect of many of the health problems that poor black people face today.

This picture can be improved. How? Two things must be done. The constituency (community) must be informed and the bureaucracy must be made more representative. These two efforts are mutually dependent. That is, a constituency that is more informed on what the bureaucracy is doing and can do is not only better off in general but is also in a better position to be involved in that bureaucracy, thus making it more representative. And, of course, when a bureaucracy is representative, it will ensure that its constituents are well informed.

Where do we begin? Very simple, the Secretary of HEW can execute an order, or if this proves improbable, the Congress can be urged to legislate a stipulation that mandates that minorities (Spanish-speaking, American Indian, Puerto Rican, or black) constitute 40% of all new appointees to advisory panels, review committees, or policy boards (in order for them to be legally constituted bodies). This would be a beginning and is currently possible; within a short period of time, minority representation would be brought to a level equal to the minority makeup of the society at large.

REFERENCES

National Academy of Sciences. Changing values and the selection process for medical school: Ethics of health care. Papers of the Conference on Health Care and Changing Values, November 27–29, 1973. Lawrence R. Tancredi (ed.). Washington, D.C.: NAS, 1974. Pp. 159–181.

U.S. Department of Health, Education, and Welfare. Minorities and women in the health fields: Applicants, students and workers. Publication #(HRA)76-22. P.H.S. Health Resources Administration. Bureau of Health Manpower. September 1975a.

U.S. Department of Health, Education, and Welfare. Public Health Service, National Institutes of Health. Public Health Service Grants and Awards Fiscal Year 1974 Funds and FY 1973 Released Funds, Part I: Research grants. DHEW Publication #(NIH)75-494, 1975b.

14

Better Health Sciences through Minority Participation: The Case for Community-Based Minority Medical Schools

ROBERT J. SCHLEGEL

The health fields occupy a special place in America. Public-opinion polls show the continuing esteem in which we hold physicians (and also bioscientists) (Gallup Poll Survey, June 1976). The average physician earned $50,000 a year in 1974 (U.S. House of Representatives, 1975). Salaries for hospital personnel have been rising relative to those of other wage earners during recent years (Davis, 1972). We as a people spent more than $118 billion dollars on health care, research, and education in 1975—about 8% of the nation's total product of goods and services. In the aggregate, the health enterprise is the nation's third largest employer (Special Analysis, 1976).

What is the record of this powerful establishment on racial justice—our continuing moral and political dilemma (Myrdal, 1944)? Are the social prestige, money, and employment being harnessed for the advancement of the nation's minorities? The most expensive elements of health care, education, and research are located preferentially in affluent neighborhoods and are unavailable to most minority people for health services, employment, or community development. Medical schools and their accompanying private and public hospitals (e.g., Veterans Administration, state university, and state psychiatric hospitals) have been migrating from the core city and its minority people. As a result, doctors in training do not learn how to provide care to underserved populations, and therefore there is an increased likelihood of further geographic maldistribution of private physicians. In addition,

ROBERT J. SCHLEGEL • Professor and Chairman of Pediatrics, Associate Dean for Public Policy, Charles Drew Postgraduate School of Medicine, Los Angeles, California.

local metropolitan hospitals in the inner city are closing down (Blake and Bodenheimer, 1974).

An increasing inequality between minority medical-student enrollment and that of others has become evident in recent years. Today, less than 7% of the undergraduate medical students are black, although American blacks constitute more than 11% of the nation's population. There was 1 white physician for every 538 white people in the United States in 1975 but only 1 black physician for every 4,100 black Americans (Sullivan, 1975). This ratio deprives black communities of health-care leadership, personnel, and role models for the young. There is likely to be a further decrease in the number of black physicians in the future as a result of the recent court decision in *Bakke vs. University of California,* a finding against preferential admissions for minorities to a medical school of the University of California (California Supreme Court, 1976).

The reasons usually advanced for the downward trend in these as well as other programs of distributive social justice is excessive fiscal cost and lack of data validating program worth, as well as the legal constitutional issue raised by the Bakke case. However, the very complexity of the health fields is also a cause of racial inequity (Richmond, 1969). I have discussed elsewhere an overall health-policy process that would guide programming to meet human and fiscal priorities, establish program worth and cost efficiency, and promote minority participation in the health fields (Schlegel, in press). The process described is based on the need for national leadership and policy to resolve current difficulties and problems in health care for all Americans, and it involves advocacy, services, education, research, and administration. However, the issue of minority access impinges on each of these global concerns and is in turn influenced by leadership and systematization involving all of them.

For example, the center at which I work in South Los Angeles maximizes minority participation by virtue of its location, its personnel policies, its citizen involvement, and its programming. Yet, it is threatened by the current fiscal crisis affecting municipal governments because its major teaching facilities are public entities of the local government. Also, while the emphasis on outreach programs by established centers in the Federal Health Manpower Act of 1976 is commendable, there is not a commensurate emphasis on the development of community-based medical schools in the legislation. The derivative consequences of current major health legislation have not been adequately examined with regard to the fallout for minority participation and the correction of ethnic inequities.

In general, policies and programs for minorities can be considered in two categories: (1) those that provide services or fiscal support and (2) those that encourage self-determination, self-help, and selfhood. The first approach transfers goods and dollars from the general public to the poor (e.g., food stamps and aid to dependent children) or provides support to care givers for free services (e.g., Medicaid). The second encourages the development of community institutions controlled by local residents. While there are obvious advantages in methods that foster independence, few examples can be found in the health fields, perhaps in part because of the highly technical nature of American medical care and in part because of protectionism on the part of the health professions (Leach, 1970).

From the perspective of this community-based medical center, minority participation is a multifaceted issue. The education of minority physicians and other health professionals is one facet. The development of a talent pool of future minority health professionals is another. The participation of all minority people in the defined population of the center in programs serving their children, their families, and themselves is equally obligatory. People can do more to improve their own health than professionals can do for them (Fuchs, 1974). It is my contention that this wide-spectrum minority participation would improve the nation's health care, its health-care system, and the health sciences. Further, it is best accomplished through the strategy of the community-based medical center.

One of the few exceptions to the prevalent mode of addressing inequities through charity is a new, community-based medical center in South Los Angeles, which is composed of two sister institutions, the Los Angeles County Martin Luther King, Jr. General Hospital (and the Southeast Health Region) and the Charles R. Drew Postgraduate Medical School. These two allied institutions represent a continuing response by local, state, and federal governments to the needs, expectations, and requests of a large minority population comprising approximately 60% of the residents of the Southeast Health Region (population, approximately 780,000). While that center is instructive as a model to study for possible national replication, it is important to analyze its weaknesses as well as its strengths and successes.

Private citizens, physicians, and dentists of the community had been actively seeking both an adequate hospital for services and continuing postgraduate education for health professionals for almost a decade prior to the Watts riots of 1966. Following the riots, the McCone Commission provided the necessary impetus to found a postgraduate medical school and a major public hospital. The hospital and the school began patient-care activities in 1972. The Los Angeles County Department of Health Services was regionalized into five entities in 1974, giving to the new medical complex the responsibility for a defined population in the Southeast Health Region. Further plans to restructure the services system (emphasizing primary care) have been largely shelved because of the department's response to the fiscal shortfall that has been felt keenly here (and across the nation) in public hospitals and health systems (Hufford, 1976).

The mission of the Drew school is to "conduct medical education and research in the context of service to a defined population so as to train persons to provide care with competence and compassion to this and other underserved populations." The Southeast Health Region operates a series of small clinics and a large, comprehensive health-care center for ambulatory patients in addition to the King hospital.

The two institutions are exceptional in several respects. First, the Drew school is the nation's only postgraduate medical school. Second, minority participation involves residents of the community (including the poor) as well as those who have attained professional status—not a unique characteristic qualitatively but quantitatively unusual. Third, the regional county and the Drew school exercise an unusual degree of control over both educational and service programs—again, not unique, but clearly contrary to present trends favoring outreach from established medical schools and central bureaucracies. Finally, the center provides leadership opportunities and managerial experience for minority administrators.

Major accomplishments of the two institutions have been:

SOUTHEAST HEALTH REGION

1. Planned and opened a $30-million hospital, a $7-million comprehensive health center, a $5-million intern and resident building, and other regional facilities.

2. Planned and began construction of a $15-million psychiatric facility and clinical-research center.
3. Recruited administrative leadership and staff (more than 90% minority composition).
4. Secured designation as the first community hospital able to admit and treat private patients in the Los Angeles public hospital system.
5. Operates a $55-million annual hospital budget.
6. Manages a $20-million annual regional budget.
7. Provides employment to 2,200 hospital workers.
8. Provides employment to regional health workers.

DREW POSTGRADUATE MEDICAL SCHOOL

1. Acquired land and temporary administrative offices.
2. Recruited a full-time faculty of 120 (80% minority professionals).
3. Developed approved residency training in most of the major specialties and recognized subspecialties.
4. Developed statements of mission and a faculty constitution as well as managerial and fiscal systems.
5. Recruited, with the King General Hospital, a resident-physician staff of 180.
6. Sought and obtained recognition as an educational entity of the State of California.
7. Constructed bioscience and community-medicine research facilities.
8. Implemented numerous health services and educational programs in the community.
9. Constructed a major child-development center.
10. Assumed responsibility for a local Headstart project numbering 13 community sites and a central kitchen.
11. Provided 189 physicians, 68% of its resident-physician graduates, to private-practice locations in the community (see Table 1).
12. Gave 255 of its resident graduates to the State of California (92% of total graduates; see Table 1).
13. Conducts clinical educational programs for 26 undergraduate medical students from other medical schools annually.
14. Developed the first medex (physician-assistant) programs in the State of California.

Table 1. Work Location of Graduates from Health-Sciences Education Programs of the Charles R. Drew Postgraduate Medical School and the Martin Luther King, Jr. General Hospital, 1973–1976[a]

	Number of graduates				Responses		Geographical work distribution		Number of graduates serving in medically underserved areas				
Program classification	73–74	74–75	75–76	Total	Not reporting	Total reporting	In-state	Out-of-state	Southeast health services region	In-state[b]	Out-of-state	Total	
Medical education (residents and interns)													
Anesthesiology	0	0	0	0	0	0	0	0	0	0	0	0	
Community medicine	0	0	2	2	0	2	2	0	2	0	0	2	
Dentistry													
Oral surgery	0	1	1	2	0	2	2	0	1	1	0	2	
Pedodontics	0	2	1	3	0	3	2	1	1	0	0	1	
General practice	5	5	3	13	0	13	0	13	8	1	0	9	
Family medicine	0	0	0	0	0	0	0	0	0	0	0	0	
Medicine	12	23	17	52	3	49	41	8	27	4	7	38	
Obstetrics/gynecology	0	4	5	9	1	8	8	0	4	2	0	6	
Pathology	0	0	0	0	0	0	0	0	0	0	0	0	
Pediatrics	10	25	4	39	4	35	27	8	14	2	4	20	
Psychiatry	1	1	3	5	3	2	2	0	2	0	0	2	
Radiology	5	4	11	20	1	19	17	2	5	4	0	9	
Surgery	0	2	7	9	3	6	5	1	4	0	1	5	
Subtotal	33	67	54	154	15	139	106	33	68	14	12	94	61%

Health sciences education (including allied health training)													
Adult nurse practitioners	0	0	8	8	1	7	7	0	4	2	0	6	
Master mental-health planners	0	24	0	24	2	22	22	0	15	3	0	18	
Medex (physician's asst.'s)	35	18	29	82	2	80	78	2	36	34	10	80	
Medical technologists	0	4	5	9	2	7	6	1	5	1	0	6	
Minority biomedical	0	1	1	2	1	1	1	0	0	1	0	1	
New careers	0	0	15	15	0	15	15	0	13	2	0	15	
Nurse–anesthetists	18	5	4	27	8	19	19	0	6	3	1	10	
Trauma nurse-specialists	0	40	12	52	1	51	50	1	30	5	0	35	
Nuclear-medicine techs.	4	3	3	10	0	10	9	1	6	1	0	7	
Radiologic technologists	0	0	12	12	3	9	9	0	6	0	0	6	
Subtotal	57	95	89	241	20	221	216	5	121	52	11	184	83%
Total	90	162	143	395	35	360	322	38	189	66	23	278	77%
Percentages under major headings	23%	41%	36%	100%	9%	91%	89%	11%	68%	24%	8%	100%	
Total percentage of reporting Drew graduates serving in medically underserved areas									53%	18%	6%	77%	

[a]Source: Office of the Registrar, Charles R. Drew Postgraduate Medical School, September 1976.
[b]Other than Southeast Health Services Region.

These achievements have resulted from the unique participation by community residents and minority professionals in the processes of care, education, and research as well as policy making. To an extent not seen in most medical teaching centers (public or private), the center is identified as an intrinsic part of the community. This community participation is reflected in the policy process of the school and proceeds in a stepwise fashion. The needs and expectations of the people are identified through personal interaction and measurement. Services are developed in response to these needs and expectations. The skills involved in the carrying out of these service tasks become the curriculum content of the medical-education programs. Research (medical, social, and behavioral) bolsters the advocacy, service, and educational components of the system. Finally, management and resource acquisition are a consequence of the mandate developed to carry out the first four steps of the policy process.

It can be seen that the community-based medical center can be highly effective in generating new health professionals for underserved areas. Further, most of those who did not choose to remain in the ghetto chose an academic career in the health sciences, another area of underrepresentation for minorities. The high retention rate of graduates of the resident-physician training program in the local community (Table 1) results from several attributes of the King–Drew center. The center shapes the attitudes, skills, and knowledge of resident physicians so that they are comfortable about practicing in the area and confident in their ability to manage its special social circumstances and medical challenges. In addition, graduates prefer to settle in an area adjacent to a center of continuing education and faculty consultation. Finally, many take pride in having helped to shape the destiny of an institution having worthy goals.

As a result of interactions between the faculty and the residents and leaders of the local community, a variety of programs have been developed to foster child-health education and development, to train community residents for new occupations (e.g., as teachers' assistants), and to provide employment for many. While criticism from community residents is constant, it is also constructive and necessary. This community has been studied for many years and by many different kinds of experts but without any prescriptive responses and serious attempts to give to the people their own means to achieve better circumstances of life through health care, educational opportunities, or local employment. The community-based center is correcting those deficiencies.

There have also been failures in the carrying out of the community mission. Our academic leadership, which follows the rational process defined above, is in many instances not a controlling or even a minor influence on program development. The public hospitals and most other elements of the service system are under a county government. The Board of Supervisors of Los Angeles County sets budgets. In the county government, the county administrative office is a more powerful influence on the budget than is the Department of Health Services. While most top positions in the Department of Health Services, the health regions, and the hospitals and the clinics are nominally in the civil service, in reality, appointments are made on political grounds by the county supervisors. The civil-service system itself has its own top-level policy makers, who dictate policies to local administrators. Hence, management and resource acquisition reflect a variety of powerful influences over which neither citizens of the community or health professionals have direct control or influence.

Because of these extrinsic influences, it is not always possible to put together programs rationally derived from our mission. Here, as elsewhere, rather than real-world priorities, empty beds and clinics become the compulsive force that generates clinical programs. Primary care is explicitly defined but not carried out because of the constraints imposed by hospital or clinic regulations made outside the local center. Of course, intrinsic limitations also play a role in our current imperfections. It is one thing to agree on broad goals, another to carry them out in daily activities with skill, energy, and commitment.

Two small federal programs account for almost all of the center's modest scientific contributions to teaching, manpower development, and the advancement of biomedical knowledge: the Minority Biomedical Support program of the National Institutes of Health and the Minority Access to Research Careers program of the National Institute of General Medical Sciences. These programs have allowed the school to develop an educational consortium with eight community and junior colleges to provide science instruction to minority students and to advance the careers of promising young clinical scientists from its own manpower pool. In addition, secondary multiplier effects accrue, since the administrative core and the scientists required for these programs are also available to guide and teach high-school students in the region and to strengthen clinical teaching in hospitals and clinics. Also, some members of the science-teaching group have been successful in obtaining research grants and contracts to carry out independent research. Yet, a larger science base is needed in the center.

Opening the school from the top down (as a postgraduate institution) rather than the bottom up (the usual developmental pattern) has certain advantages. It finesses traditional conflicts between basic scientists and clinical faculty members, for example. However, certain disadvantages result from the absence of basic science departments. It is more difficult to carry out good clinical science in the absence of such cross-cutting disciplines as biochemistry, genetics, and physiology, among others. Clinical scientists are more difficult to recruit. Clinical teaching and services are weakened when they are not subjected to critical scrutiny and constant reinvigoration by a strong local biomedical-science establishment. This lack will not be corrected until the Drew school develops its own undergraduate medical school and basic science departments.

Clearly, the new medical center is in a position to contribute to medical education through the development of competency-based learning programs that respond to specified priorities of service. This response has particular merit when applied to the measured needs and expectations of underserved minorities, who have high mortality and morbidity rates (U.S. Department of Health, Education, and Welfare, 1975) and account for most of the excess deaths and preventable diseases in our service region and in the United States generally, compared to other industrialized nations. Further, the center provides unusual opportunities for academic as well as community leadership by minority professionals. It should also help to reverse the current downward trend in the proportion of minorities among medical-school graduates.

However, there is a strange reluctance on the part of some of our political and administrative leaders to accept the identification of the Drew school or the King hospital as a minority institution. Clearly, the State of California needs to educate more minority physicians (Table 2). Yet it has been stated by state policy-makers and by health policy-makers in the federal government that programs or institutions designated for minorities will fail to compel broad-based interest and support among Americans in the current political and social climate. For

Table 2. Demographic Characteristics of First-Year Students Selected by Medical Schools in the State of California, 1974[a]

Medical schools	Number of students	Men	Women	Black American	Mexican-American	Mainland Puerto Rican	American Indian
Public:							
University of California at Davis	100	66	34	5	9	0	1
University of California at Irvine	79	58	21	17	12	1	1
University of California at Los Angeles	147	123	24	6	16	1	2
University of California at San Diego	96	75	21	5	4	0	0
University of California at San Francisco	147	95	52	14	13	1	1
Total	569	417	152	47	54	3	5
Private:							
Loma Linda University	154	125	29	9	2	0	1
University of Southern California	130	99	31	5	9	0	1
Stanford University	94	65	29	11	8	0	1
Total	378	289	89	25	19	0	3
Grand Total	947	706	241	72	73	3	8
Percentage	(100%)	(75%)	(25%)	(8%)	(8%)	(0.3%)	(0.8%)

[a]Phase I Report Educational Policy and Curriculum Committee, Charles R. Drew Postgraduate Medical School, Chairman, George Locke, M.D.

that reason, we can expect difficulty in obtaining the support and the recognition of the proposed extension of our current activities to undergraduate medical education, an essential evolutionary step if we are to maintain high standards of service and education, as well as community responsibility. In addition, the Association of American Medical Colleges has recently issued a position paper discouraging the development of new affiliated two-year clinical-medical schools. The Carnegie Commission has published a study indicating that the nation now educates too many physicians (although it distributes them poorly by specialty and by geographic region). Altogether, these trends are not propitious for the development of any new medical school, especially one designated for broad-based minority participation.

It can be seen that an unusual range of services is likely to occur in the community-based medical center. Services respond to need rather than professional tradition. Hence, collective efforts are likely to occur among health professionals, citizens, and the members of other professions. These efforts result in part from the explicit identification of the things that people can do for themselves and their own health, those that can be done by individual health professionals, and those that require teamwork between health centers and other community agencies (often negated by considerations of professional status and territoriality). In the community-based medical center, the focus is on getting the job done rather than on professional preconceptions and prejudices. Of course, the actual fact is that at King–Drew, the goals and objectives together are only one influence on attitudes and behavior. Newcomers among the professional staff invariably go through a difficult introductory period. When the center first opened, we established a series of primary-care teams as the nexus of services and postgraduate training in one department. The system failed because an insufficient number of the faculty were expert in primary care (or committed to it) and because of lack of accountability among the resident-physician staff. However, the center policy does stimulate directed movement and a reorientation of professional practice toward the real-world priorities of health care.

Many attempts have been made in recent years to increase minority participation in the health fields and the biosciences. Most have taken the form of special programs to increase access to health care or enrollment in professional education programs. Many of these efforts appear to be losing their vitality, with a resultant further increase in the inequity of services and of opportunities for technical and professional education.

In view of these less-than-satisfactory results, it seems worthwhile to examine alternative approaches to increasing minority participation in health fields. One alternative is the creation of community-based minority institutions capable of conducting both services and education. This approach is not new in America nor in other nations. However, it is worthwhile to examine those unusual instances in which it has been applied to the problems of distributive social justice. In this regard, the King–Drew medical center is suitable for analysis.

The King–Drew medical center is unusual in several respects. It is the nation's only postgraduate medical school. It is historically and currently a result of broad-based participation by minority peoples. All citizens of the defined population can interact with the center through its programs of education and service. The leadership of the school, the hospital, and the health region is a minority leadership. The majority of faculty, staff, postgraduate trainees, and students are themselves derived from the minorities they serve. This new center is located in an underserved region of poor minority people.

Judging from its record to date, a movement toward establishing community-based medical centers would make possible minority participation in a broad scale. The results would help in the economic revival of poor communities. New jobs would be created. Preprofessional programs of manpower development would result. A reality base would be provided for health education and health promotion. Health services would become more accessible. The distribution of health professionals would be improved because of the presence of postgraduate training programs specifically designed for underserved communities and because of the continuing presence of a major service, education, and professional center for private health practitioners. These are some of the many advantages of this approach as revealed in the examination of the accomplishments of the King–Drew medical center.

Drawbacks and potential pitfalls are also evident in the brief history of that center. It is by no means certain that any major center so heavily dependent on funding by a local government is viable. Serious problems are encountered in attempts to move the academic establishment to accept or to foster any new medical school, much less a two-year affiliated school or a minority school. Without such acceptance or encouragement, it is difficult indeed to launch an enterprise of such complexity as a modern health-education center.

There is also the question of the long-term viability of a postgraduate school. It is not feasible to develop or sustain the necessary high-quality science programs in the absence of basic science departments. Postgraduate training and faculty development are adversely affected by the lack of strong programs in the health sciences. Only a conscious policy of fostering community-based medical centers can give the King–Drew program a destiny. Such a policy would advance minority participation in the health fields. More importantly, it would serve all of the nation's minority peoples.

The strategy of the community-based medical center is presented as one way to increase minority participation in the health sciences. However, it does something more. It provides a structural arrangement that will strengthen the health sciences. It encourages science programming for the young people of minority communities, increasing the talent pool for the biomedical-science and health professions through education and career guidance. There is also created a natural interface between minority professionals and the target minority communities, enhancing informed advocacy and strengthening the quality of community life.

REFERENCES

Blake, E., and Bodenheimer, T. Hospitals for sale. *Ramparts*, 1974, *27*.

California Supreme Court. *Bakke vs. the Regents of the University of California.* San Francisco 23311, San Francisco, California, 1976.

Davis, K. *Economic theories of behavior in nonprofit, private hospitals.* Washington, D.C.: Brookings, 1972.

Fuchs, V. *Who shall live?* New York: Basic Books, 1974.

Gallup Poll Survey, June 1976.

Hufford, H. *County of Los Angeles Digest*, 1976, *9*, 8.

Leach, G. *The biocrats.* Philadelphia: McGraw-Hill, 1970.

Myrdal, G. *An American dilemma.* New York: Harper and Row, 1944.

Richmond, J. B. *Currents in American medicine.* Cambridge, Mass.: Harvard University Press, 1969.

Schlegel, R. J. How to think about health policy. In H. Hugh Fudenberg and Vijaya L. Melnick (Eds.), *Biomedical sciences and public responsibility*. New York: Plenum Press, in press.

Special Analysis. Budget of the U.S. government fiscal year 1977. Washington, D.C., 1976.

Sullivan, L. W. Testimony before the Subcommittee on Health, Committee on Labor and Public Welfare, U.S. Senate, 1975.

U.S. Department of Health, Education, and Welfare. Health in the United States, #DHEW (HRA) 76-1233. Washington, D.C., 1975.

U.S. House of Representatives. Hearings before the Subcommittee on Health and Environment, 1975.

IV. Financial Support for Minority Scientific Activities in Education and Research

15

Funding of Minority Programs from the Private Sector: A Perspective from the Josiah Macy, Jr. Foundation

MAXINE BLEICH

I am honored to have been asked to share with you some of my observations and thoughts about an area of vital national concern: the increase of minority representation in medicine and the related health professions.

Minority-student representation in the nation's medical schools has increased in the last decade. In 1966, there were 266 black first-year medical students, two-thirds of whom were enrolled in the country's two black medical schools, Howard and Meharry. Today there are 1,036 black first-year students, the vast majority of whom are enrolled in the nation's medical schools other than Howard and Meharry.

Although minority enrollment has increased four times over the past 10 years, this encouraging statistic is offset by the fact that this past year, 14.4% of the black freshmen repeated courses in the first year as compared with 1.2% of the majority freshmen (*Journal of the American Medical Association,* 1975). Attrition data, which I will present later, further demonstrate that the minority students as a group enter medical school with a statistically demonstrable educational handicap.

The high rate of repeats in the first year has accompanied the increased numbers of minority students. When this is taken into consideration, we can see that the increase in newly admitted first-year black medical students actually began to plateau in 1971–1972, and for this academic year, we saw a decrease in the number of black first-year medical students—from 1,106 in 1974 to 1,036 in 1975. There are many

MAXINE BLEICH • Associate Program Director, Josiah Macy, Jr. Foundation, New York, New York.

indications to lead one to believe that this decline in admissions will continue.

A measure of integration of the nation's medical schools has been achieved during the past decade. However, an equally great challenge remains: the appropriate academic preparation of minority students prior to entrance to medical school. Our public schools, during these 10 years, have demonstrated a diminished capacity to provide the knowledge and the discipline in reading and mathematics and to bring about the development of the quantitative abilities necessary to a competitive education. As a consequence of inappropriate preparation in public school and often in college, a high percentage of minority students are not successful as undergraduate premedical majors. Many of those who are successful find it necessary to repeat courses in medical school.

The programs I shall describe have benefited individual students, schools, and organizations, but they have not affected the needed structural change in the public schools. The educational issues unresolved by the nation's public schools remain the greatest challenge, along with their social, economic, and political implications.

The Josiah Macy, Jr. Foundation is a private, philanthropic foundation whose concern is medical education. In the mid-1960s the directors of the foundation established a program to increase the opportunities for members of minority groups to study medicine and the related health professions. We have included American blacks, mainland Puerto Ricans, Mexican-Americans, and American Indians. The foundation awards approximately $1.5 million a year; over a 10-year period close to 50% of its grants were awarded in support of these programs.

In the development and implementation of programs, the Macy Foundation is able to draw upon the special knowledge of experts in a particular field. In addition to using the services of these consultants, the foundation sponsors a conference program, which over the years has brought together informed minority educators and students to discuss the issues. These two mechanisms have provided an opportunity for minority consultation at all levels in the development and implementation of foundation-sponsored programs.

The first Macy program to assist minorities to enter the health professions was established in 1966 under the direction of Dr. William E. Cadbury, Jr., then Dean of Haverford College. It was designed to demonstrate to the medical schools that black students, particularly those who had graduated from traditionally black colleges, could be successful candidates for medical training.

Table 1. Total U.S. Minority-Student Enrollments, 1969–1970 through 1975–1976[a]

Group	1969–70 Number	%[b]	1970–71 Number	%[b]	1971–72 Number	%[b]	1972–73 Number	%[b]	1973–74 Number	%[b]	1974–75 Number	%[b]	1975–76 Number	%[b]
lack (Americans)	1,042	2.8	1,509	3.8	2,055	4.7	2,582	5.4	3,045	6.0	3,355	6.3	3,456	6.2
merican Indian	18	[c]	18	[c]	42	.1	69	.2	97	.2	159	.3	172	.3
Mexican-American	92	.2	148	.4	252	.6	361	.8	496	1.0	638	1.2	699	1.3
uerto Rican (mainland)	26	[c]	48	.1	76	.2	90	.2	123	.2	172	.3	197	.4
Total	1,178	3.0	1,723	4.3	2,425	5.6	3,102	6.6	3,761	7.4	4,324	8.1	4,524	8.1

Sources: W. F. Dube, U.S. medical school enrollments, 1969–70 through 1973–74, *Journal of Medical Education,* March 1974, *49,* 304–306; Association of American Medical Colleges, 1974 fall enrollment questionnaires, *The Advisor,* April 1974; 1975–1976 data from the Association of American Medical Colleges.

Percentage of Total Enrollment

Less than 0.1%

Table 2. First-Year Black-Student Enrollments, 1969–1970 through 1975–1976[a]

Group	1969–70 Number	%[b]	1970–71 Number	%[b]	1971–72 Number	%[b]	1972–73 Number	%[b]	1973–74 Number	%[b]	1974–75 Number	%[b]	1975–76 Number	%[b]
Black (Americans)	440	4.2	697	6.1	882	7.1	957	7.0	1,023	7.2	1,106	7.5	1,036	6.8
American Indian	7	.1	11	.1	23	.2	34	.3	44	.3	71	.5	60	.4
Mexican-American	44	.4	73	.6	118	1.0	137	1.0	174	1.2	227	1.5	224	1.5
Puerto Rican (Mainland)	10	.1	27	.2	40	.3	44	.3	56	.4	69	.5	71	.5
Total	501	4.8	808	7.0	1,063	8.6	1,172	8.6	1,297	9.2	1,473	10.0	1,391	9.1

[a]Sources: W. F. Dube, U.S. medical school enrollments, 1969–70 through 1973–74, *Journal of Medical Education,* March 1974, *49,* 304–306; Association of American Medical Colleges, 1974 fall enrollment questionnaires, *The Advisor,* April 1974; 1975–76 data from the Association of American Medical Colleges.

[b]Percentage of total enrollment.

Table 3. Students Admitted to U.S. Medical Schools, 1971–1972 through 1974–1975, and Percentages Still in School, June 1974 and June 1975[a]

	Admitted 1971–72	Retained June 1974 Number	%[b]	Admitted 1972–73	Retained June 1974 Number	%[b]	Admitted 1973–74	Retained June 1974 Number	%[b]	Admitted 1974–75	Retained June 1975 Number	%[b]
Black (American)	758	649	86	810	716	88	864	802	93	934	886	95
American Indian	21	21	100	31	30	97	37	37	100	63	62	95
Mexican-American	117	110	94	138	133	96	169	166	98	203	198	97
Puerto Rican (mainland)	33	30	91	37	34	92	48	47	98	60	59	98
All students	10,962	10,500	96	12,520	12,118	97	13,062	12,842	98	12,892	12,750	99

Source: Medical education in the United States, 1973–1974, supplement to *Journal of the American Medical Association,* January 1975, 231; Medical education in the United States, 1974–75, supplement to the *Journal of the American Medical Association,* December 1975, 234.

Percentage of total enrollment.

Each spring for five years, Dr. Cadbury visited the black colleges to identify potential medical-school applicants. Most of the students that Dr. Cadbury selected wanted to study medicine but either had not applied to medical school or had not been accepted. As participants in the program, the students enrolled in a special summer session at Haverford, in Pennsylvania, and in later years at Oberlin, in Ohio. Following the summer program, the students enrolled for a full academic year in one of seven participating colleges: Bryn Mawr, Haverford, Kalamazoo, Knox, Oberlin, Pomona, or Swarthmore. According to their particular needs, the students enrolled in regularly scheduled courses in chemistry, biology, mathematics, and English.

During the five years of the program, 95% of the 76 students completed the year and a half; more than 90% of these students were accepted and enrolled in medical school. These students were indeed pioneers, the first group of minority students to begin to integrate the nation's medical schools.

In order to increase the number of minority-group medical students, the medical schools had to accept them, and, of course, the complementary recruitment, preparatory, and retention programs had to be developed.

To stimulate the necessary activity, in 1968 and 1969 the foundation sponsored regional conferences to bring together the premedical advisers from the black colleges with the medical-school admissions officers in the same area. These conferences provided an opportunity for each of the two groups to get to know one another and to discuss the key issues relating to the preparation, the enrollment, and the retention of minority medical students. One of the major developments was that for the first time the Medical College Admissions Test (MCAT) was nationally exposed as a mechanism that excluded minority-group students with potential from being accepted to medical school. Because of the difference in scores obtained on the MCAT by minority and majority candidates, the medical schools have learned to broaden their admission criteria to obtain a better analysis of a minority student's potential and ability.

The foundation also established a program to help the medical schools initiate their efforts to recruit, prepare, enroll, counsel, and retain an increased number of minority students. Between 1967 and 1975, the foundation has supported programs for minority students in more than 45 of the nation's 114 medical schools. Funds were provided for direct recruitment activities; for programs that brought minority high-school students to the academic medical center to learn of its many opportunities; for programs that provided course work in the basic sciences, mathematics, and reading, and exposure to the laboratories and clinical activities of the academic medical center for minority college and entering freshmen and medical students; and for the establishment and support of offices of minority affairs. These programs affected students at almost every level of academic preparation and provided us with a wealth of information.

Programs directed toward high-school students have reflected the understanding that youngsters need to be motivated and made aware of opportunities in the health professions at an early age. Many medical schools have provided work in their laboratories as well as special courses in science and mathematics for high-school students. The programs that have a significant number of their graduates in four-year colleges and/or enrolled in professional schools demonstrate that these special academic experiences, along with careful academic and personal counseling and tutorial and financial assistance, are successful in preparing the students for college and a professional education. The most successful programs have had full-time leadership to provide the necessary tutorial and counseling assistance. However, for the most part, these programs have not had the cooperation of the public-school system and, as a consequence, have not been successful in strengthening the academic courses and guidance-counseling services provided by the public schools.

Many medical schools have provided similar types of programs for college-level students. The successful programs have provided rigorous course work in the sciences, mathematics, and English. Also, reading, studying, note-taking, and test-taking techniques have been taught, and the students have learned to use the library. Often, the programs have included laboratory work and clinical exposure. In addition, they have provided guidance and counseling on an individual basis and seminars relating to the many opportunities in the health professions as well as to the health concerns of minority communities. Many medical schools look favorably upon applicants who have participated in these programs. However, as is demonstrated by the high rate of academic difficulty experienced by many minority students, these programs alone cannot overcome all of the problems. They do, however, provide an opportunity for the faculty of the medical schools to develop working relationships with the faculty of the colleges in which the students are enrolled. These relationships must be aggressively pursued by both the medical schools and the colleges in order to assure the continuation of a rigorous education at the undergraduate level in preparation for medical school.

In an effort to reduce some of the tensions for first-year medical students whose entering records suggest that they might have more academic problems than other students, some medical schools have established summer programs that provide rigorous course work in biochemistry and physiology, as well as exercises in test taking, reading, and note taking.

In addition, many medical schools have established offices of minority affairs. These offices are responsible for the easing and, it is hoped, the resolution of many of the underlying problems of students from minority groups, such as inadequate high-school or college preparation; the students' ambivalent attitudes relating to their academic preparation and their role in medical school; financial problems; and feelings of isolation from classmates and faculty members. These offices coordinate tutorial assistance, guidance and counseling, financial aid, and direct recruitment activities.

History speaks for itself! The number of minority-group students enrolled in the freshman year in United States medical schools has increased: from 292 in 1968 to 1,391 in 1975. Between the years 1969 and 1971, approximately 70% of the minority applicants were accepted. This was most impressive, for at that time, less than 50% of the total applicants were accepted. However, since 1973, approximately 40% of the minority applicants have been accepted, which is comparable to the acceptance rate of all students. From available information, it appears that the pool of qualified minority-group students has not grown at the same rate as the pool of all students. The diminished size of the pool as well as the high rate of academic difficulties experienced by minority-group students and their extensive financial needs has resulted in a real decline in their admission for 1975–1976.

BLACK COLLEGE PROGRAMS

In response to the indications that the pool of minority candidates was reaching a plateau and that the incidence of academic problems continued to be an important issue for these students, the foundation turned its attention to undergraduate preparation; perhaps we could be effective in strengthening programs prior to entrance into medical school.

In 1970, the directors of the foundation established a major program directed at the black colleges. This decision was made in part because more than 80% of the black medical students were graduates of these colleges, and, of course, the schools had an unswerving dedication to the preparation of black Americans for professional leadership. Initially, the program was designed to strengthen health-profession advisory services. One aspect included a month-long summer institute for health-profession advisers from selected black colleges. The advisers, for the most part, were chairmen of the biology or chemistry departments.

In the institutes, the advisers learned firsthand about opportunities in the health professions; the type of academic preparation needed to equip candidates to enter and remain in professional schools; and the academic, financial, social, and emotional experiences of minority-group medical students. They also took the MCAT and modified versions of the Dental Aptitude Test (DAT) and the Graduate Record Examination (GRE) and met with representatives of the testing centers.

These institutes were supplemented by direct grants made to 13 black colleges. The grants were used to support a portion of the advisers' salaries, tutorial programs, the activities of the health-professions societies, and the direct recruitment of high-school students for the premedical program at the college.

From our close association with the black colleges and many of their graduates enrolled in medical school, we learned that although the colleges provided a fine education, much of the biology curriculum was not modern biology. For many complex social and economic reasons, many of the biology departments had not been able to offer courses in the modern molecular and cellular biology that has developed over the past 20 years. Hence, many black college graduates were at a disadvantage in professional schools, competing with their white classmates who had received training in modern biology.

In response to this problem, the foundation directed its efforts to the development of biology faculties at selected black colleges. In 1975, the foundation held the first summer institute, "Premedical Education: Biology," at Atlanta University. Letters were sent to the presidents of 42 black colleges, inviting each of them to nominate a faculty member to participate in the institute. Those enrolled were selected according to the following guidelines: they had not received the Ph.D. degree in molecular biology within the last two years, since the information offered by the institute would be repetitive; they had expressed or demonstrated an interest in teaching quantitative-biology courses as part of the undergraduate curriculum; and they were permanent members of the faculty above the rank of instructor. Participants from 30 colleges were selected and 29 enrolled.

This summer a second institute at a more advanced level, "Premedical Education: Biology II," will be held at Atlanta University. Representatives from 15 schools that were considered to be in a position to move ahead quickly in modern biology have been invited for a more advanced course. The selection of the 15 schools merely represents the practical limits of the program and should by no means be interpreted as a decision that other black colleges do not have the potential to develop modern biology curricula.

As a follow-up to the 1975 institute and in preparation for the 1976 program, the codirectors of the institute and I are visiting each of the 15 colleges that will be represented this summer. We are meeting with members of the departments of biology, chemistry, mathematics, and physics, as well as with administrations, to learn firsthand about the academic programs offered to students preparing for medicine and dentistry or graduate work in the basic sciences. We hope that these visits will provide information that will benefit all of us who are helping the schools modernize their science programs.

In addition to this faculty development, grants have been made to four black colleges to help strengthen the students' preparation for medical and graduate schools. Funds are supporting summer programs for entering college freshmen. These programs introduce the students to principles in biology and chemistry and provide special exercises in mathematics and reading. These funds also provide tutorial services during the academic year and formal review courses—some for credit—in preparation for the MCAT, the DAT, and the GRE.

Our most recent initiative has been an award to the Marine Biological Laboratory at Woods Hole, Massachusetts. The MBL will develop a program to train blacks and members of other minority groups in modern biology and biomedicine, with special attention to be given to the needs of younger faculty members and students in selected black colleges. The participants will be at the pre- and postdoctoral levels. The program will be coordinated with the MBL's "Steps toward Independence" program, which provides scholarships, equipment, and a wide range of essential research services for nationally selected junior investigators and students.

The programs at the college level provide an excellent opportunity to learn the extent and the specific nature of the academic deficiencies of the students, most of which center around mathematics and reading. If these problems can be confronted appropriately, we can then expand the pool of well-prepared minority students for medical, dental, and graduate schools and perhaps begin to reverse the trend of inappropriate preparation for college and graduate or professional school.

Much remains to be done that can be accomplished at both the college and the medical-school levels to provide the academic reinforcement that is needed by many minority students. In addition, the greatest challenge remains: to help the nation's public-school systems provide their minority students, as well as all students, with an education that will prepare them for college and professional school.

REFERENCE

The Journal of the American Medical Association, 1975, *234*(13), 1339.

16

Graduate Fellowship Opportunities for Minorities Most Commonly Discriminated Against in Higher Education

SAMUEL M. NABRIT

The Council of Southern Universities is a group of universities organized in 1952 for the purpose of upgrading the faculties of southern institutions. Its members are Duke University, Emory University, the University of North Carolina, Rice University, the University of Texas, Tulane University, Vanderbilt University, Louisiana State University, and the University of Virginia. The council was funded in its early years by grants from the General Education Board, an organization established by John D. Rockefeller to administer his educational philanthropies. These grants enabled southern universities to gain parity with institutions in other regions with respect to the number of faculty members with doctorate degrees.

In 1954, the late Robert L. Lester, former secretary to the Carnegie Corporation, spearheaded the establishment of the Southern Fellowships Fund. The fund was established as an administrative agency of the Council of Southern Universities and supported the graduate training of students as potential faculty members for southern institutions. Once the goal of upgrading the faculties of the major institutions was completed, the council, in concert with the 10 black colleges, sought additional funds for the primary purpose of strengthening the faculties of the predominantly black colleges. Since 1966, the Southern Fellowships Fund, the operating agency of the council, has provided graduate

SAMUEL M. NABRIT • Executive Director, The Southern Fellowship Fund, Atlanta, Georgia.

fellowships to present and potential faculty members of these institutions without racial or sex discrimination. The funding for this phase of the council's activity has been provided by the Danforth and Andrew W. Mellon foundations with assists from CPC, Inc. and from the Equitable Life Assurance Society.

Of the several hundred fellows that have been supported by the fund, 87 have received training in the sciences. Renewals were available for these 87 students to the completion of their studies, and they were encouraged to obtain the doctorate. Some students completed doctorates in the teaching of science and in science education, but the majority of the students earned the terminal degree in their chosen field of science with the possible exception of mathematics. Mathematics is exceptional in that many students complete the course work in mathematics, but write the dissertation in the teaching of mathematics, thereby getting the degree in education rather than in mathematics.

The Ford Foundation launched a doctoral program for minority students patterned after the program of the Southern Fellowships Fund. It differed in that it was launched at a time when broader opportunities existed for minorities to move to the major universities as faculty members, under the stimulus of "affirmative action." The demand for black studies, as well as the pressures exerted by the newly enrolled black students, created a temporary necessity for more black faculty.

In the program sponsored by the Ford Foundation, graduate fellowships for black Americans, American Indians, Puerto Ricans, and Mexican-Americans were supported initially as an in-house operation, beginning in 1969. Recently, however, the program for black Americans was transferred to the Council of Southern Universities, while the other three programs are administered by the Educational Testing Service.

The first two years of the fellowships for black Americans provided 250 awards each year. In addition to the renewal of some of these 500 fellowships, we have awarded 140 new fellowships for the third year of the program. Since 1970, 76 awards have been made to students in the sciences.

The 1976–1977 academic year is the last year for cohort sets from the senior class to be admitted to support. They will be funded up to five years, and the present program will phase out by 1981. A few advanced graduate students will be admitted each year until 1980, but funding will terminate for all awardees in 1981 under the terms of our present grant. The grants in both of these programs average $6,300, including tuition and fees. Of that amount, $300 per month is the basic stipend for the graduate fellowships for black Americans, and the Southern Fellowships Fund's stipends range from $2,800 to $5,100. An award not exceeding $1,000 is available to fellows to assist in research and dissertation preparation. Approximately 75% of the fellows of the Southern Fellowships Fund have obtained employment in black institutions; the other 25% are divided among white institutions, industry, and government, with the majority going to white institutions.

The Danforth Graduate Fellowships enroll students who contemplate graduate careers in the sciences and other areas. This support is available for four years of study. The revised Danforth program will assure minorities of 25 fellowships each year out of a possible 100 awards.

ication and fees. Of that amount, $300 per month is the basic stipend for the graduate fellowships for black Americans, and the Southern Fellow-

[illegible] fellows consists of research and dissertation preparation. Approximately 75% of the fellows of the Southern Fellowships Fund have obtained employment in black institutions; the other 25% are divided among white institutions, industry, and government, with the majority going to white institutions.

The Danforth Graduate Fellowships enroll students who contemplate graduate careers in the sciences and other areas. The support is available for four years of study. The revised Danforth program will make a minimum of 25 fellowships each year for a possible 100 awards.

17

National Institutes of Health Minority Research and Training Programs

ZORA J. GRIFFO

It is a pleasure to participate in this symposium and to discuss with you the National Institutes of Health and the NIH minority programs.

In historic terms, the NIH goes back no more than a quarter of a century, its minority programs no more than half a decade. Until 1948, there was only one categorical institute—the cancer institute—in addition to the small National Institute of Health. Then Congress authorized two new institutes to explore heart and dental diseases and NIH was pluralized to the National Institutes of Health. After 10 years of near stability, the Russian *Sputnik* ushered in a decade of unparalleled growth and expansion. From 1956 to 1967, the NIH annual appropriation increased from $98,458,000 to $1,412,983,000. Commensurate with this growth, the NIH became the world's leading biomedical-research agency in the fight against mankind's debility and disease.

At present, we are in another era of stability. In constant dollars, the NIH budget has increased only 20% since 1967. Even more significantly, this increase does not reflect a corresponding growth of all the NIH institutes, but mainly the National Cancer Institute (NCI) and the National Heart and Lung Institute (NHLI).

At present, the NIH is a mosaic of bureaus, institutes, and divisions, each having its own disease-oriented or biomedical-science support mission. The total NIH mission is to advance the health and well-being of man through the support of biomedical research, research training, research resources, and communications.

The NIH accomplishes this mission in two principal ways. First, it

ZORA J. GRIFFO • Special Programs Officer, Office of the Director, National Institutes of Health, Bethesda, Maryland.

conducts research in its own extensive facilities in Bethesda, Maryland. Second, through grants and contracts, it supports the work of nonfederal investigators at various medical schools, universities, and hospitals throughout the country. Through this approach, a strong partnership has evolved between the NIH and the grantee institutions, which have now collectively become a national resource combining education and research. The hallmark of this national resource is its size, its diversity, and its excellence.

Ultimately, the strength of this national resource rests in the large manpower pool of doctoral scientists. Between 1938 and 1972, the NIH helped to train approximately 94,000 scientists. Of these scientists, 60% are Ph.D.'s and most of the rest are M.D.'s.

As for minority scientists, precise figures are not available. It can be stated, however, without hesitation, that the number is very small. The reasons are too complex to be analyzed here. Very prominent, nonetheless, is the fact that there has been a consistent dearth of minority college graduates prepared and motivated to compete for NIH training support. To correct this situation, the NIH has made a decision to go beyond its mainstream activities and to make a special effort to promote the entry of talented minority individuals into biomedical science.

As we know from recent statistics, 62% of the black M.D.'s and 72% of the black Ph.D.'s in science have received their undergraduate education at black institutions. Thus, strengthening these schools is one of the most direct measures the NIH could take to increase the potential pool of minority scientists. In February 1971, the President encouraged government at all levels to help black institutions compete with other colleges and universities for students and faculty. That same year, the Senate Appropriations Committee authorized the Division of Research Resources, NIH, to develop a program for strengthening health sciences at predominantly black colleges.

After extensive consultations with the minority community, the division responded by launching the Minority Schools Biomedical Support Program—later changed to the Minority Biomedical Support Program (MBS). Second, and also after a broad dialogue with minority scientists and administrators, the National Institute of General Medical Sciences proposed to set aside funds from its fiscal year 1972 budget to strengthen minority schools through faculty fellowships and visiting-scientist awards. Congress responded by creating the Minorities Access to Research Careers (MARC) Program.

Both programs are unique for the NIH. The NIH ordinarily supports those projects, activities, and individuals that are already at the forefront of biomedical progress. By contrast, the MBS and MARC programs address the more preliminary aspects of this continuum.

Early eligibility criteria for both programs referred to "predominantly minority institutions" requiring 50% or more enrollment of ethnic-minority students. On June 30, 1975, new regulations went into effect for the MBS program. These will allow the program to fund also (1) four-year institutions that have significant minority enrollments but not necessarily over 50%, provided there is a past history of encouragement and assistance to minorities; (2) two-year colleges with a minimum of 50% minority enrollments; and (3) American Indian tribal councils that perform substantial governmental functions or an Alaska regional corporation as defined in the Alaska Native Claims Settlement Act. Similarly, the current eligibility language for the MARC program also refers to "substantial numbers of minority students."

Under its guidelines, the MBS program provides release time for faculty from teaching duties to engage in research and also financial support for students who collaborate on these projects. Equipment, supplies, travel, and consultants are likewise made available. The MARC program supports the doctoral and postdoctoral training of minority-school faculty at recognized centers of excellence with the expectation that they will return to the sponsoring institutions after the completion of their training. In addition, the program enables outstanding scholars to join minority institutions for a limited period of time to enhance research in their areas of competence.

To date, the history of both programs is highly encouraging. From a modest budget of $2 million in fiscal year 1972, the MBS program grew to $5 million in FY 1973, $8 million (includes $1 million of impounded FY 1973 funds) in FY 1974, and $7,662,964 (includes $341,964 from NCI and NHLI for support of individual disease-oriented projects) in 1975. MBS funds are appropriated by Congress. The funds for the MARC program have remained at $5 million; they are set aside annually from the institute's budget.

To date, the MBS and MARC programs are to be credited with substantial gains. In less than one year from its inception in 1971, the MBS program made awards to 38 or to roughly one-third of the traditionally black colleges and universities. By FY 1975, this number more than doubled to include 74 grantee institutions. Faculty participation grew from 199 in FY 1972 to 589 in 1975. The number of students receiving MBS support rose from 288 to 1,008 for undergraduates and from 44 to 184 for graduate students. Similarly, the MARC program now supports 350 graduate fellows in 67 eligible institutions.

In addition, significant future growth may be anticipated for the NIH minority programs because of the expressed intent of several disease-oriented institutes, particularly the NHLI and the NCI, to play a role in NIH minority programs.

The prospect of an all-institute participation posed a challenge to the NIH. If unguided, it could lead to a conglomerate of separate efforts with narrow needs, disparate goals, and splintered resources; gap areas might appear on the one hand and overlaps on the other. Recipient institutions might meet with inordinate administrative burdens for relatively minor returns. Thus, it became incumbent for the NIH to develop a cohesive approach to minority programs.

The present system was introduced in 1974. It not only extends horizontally but also stratifies vertically within the NIH hierarchy. The overall leadership for the program rests in the Office of Extramural Research and Training in the immediate Office of the Director, NIH. It is supported, in an advisory capacity, by the Coordinating Committee for NIH Minority Research and Training, composed of representatives from all institutes and divisions of the NIH.

On the operational level, the MBS and MARC programs and the Minority Program in the Neurosciences (MPN) of the National Institute of Neurological and Communicative Disorders and Stroke are the three independent program entities of the NIH. Furthermore, the MBS and MARC programs have now also become conduits for the disease-oriented institutes to support minority schools. Such support is provided for categorical research congruent with each institute's mission on a project-by-project basis. According to the new system, the MBS and MARC program staff remain fully responsible for the total management of grants in their programs. They identify projects with disease-oriented objectives and initiate cooperative agreements with the appropriate institutes. Institute staff, on the other hand, exert strong leadership in developing categorical research at minority institutions. They also evaluate the progress made on projects supported by their institutes.

The new approach has already been tested on a limited basis and is judged to be effective. In FY 1975, over $340,000 were transferred to the MBS program from the NCI and the NHLI. A similar agreement has been reached between the NCI and the MARC program. The National Institute of Allergy and Infectious Diseases is currently negotiating similar agreements. We expect that a number of the other institutes will follow suit.

Thus, instead of a collection of independent efforts, this system offers a cohesive approach as well as a framework for significant future expansion. While still untried on a larger scale, the potential for growth is deemed to be excellent. The growth will depend on the ability of the categorical institutes to support cooperative projects. Even more importantly, it will depend on the level of research activity at minority schools and, ultimately, on the number and quality of proposals submitted to the NIH for funding. In turn, the NIH will need to remain attuned to the changing environment; build on the advice from the community it aims to serve; and be continually prepared to improve the system.

In conclusion, I should like to emphasize that the programs described here at some length are merely an entry point into the biomedical-research arena. All health scientists, minorities and others, are to be strongly encouraged to compete as early as possible through the regular NIH mechanisms. This will be the ultimate measure of becoming part of one of the finest of human endeavors.

18

Minority Programs Funded by the National Science Foundation

JAMES W. MAYO

The National Science Foundation (NSF) is an agency of the federal government that was established in 1950 to advance scientific growth and development in the United States. The foundation fulfills this responsibility primarily by sponsoring scientific research and training in academia, encouraging and supporting improvements in science education, and fostering the exchange of scientific information. The foundation supports scientific research and education projects in the mathematical, physical, medical, biological, social, and engineering sciences.

The development of programs for minorities at the NSF resulted from some independent efforts at the agency and were a part of an overall effort of the federal government to increase assistance to black colleges.

In 1966, a science-development symposium dealing with the special problems of developing better science programs at institutions with predominantly black enrollments was sponsored by the foundation. Following this symposium, there were staff papers, analyses of ongoing programs, evaluation of possible future activities, and other program-development activities, all of which were geared toward the foundation's assisting the development of science programs at black institutions.

In 1969, a meeting was held on "Assistance to Developing Institutions" and an Advisory Committee on Developing Colleges was formed. The committee was comprised of representatives from profes-

JAMES W. MAYO • Deputy Division Director, Division of Science Education Resources Improvement, Directorate for Science Education, National Science Foundation, Washington, D.C.

sional societies of biology, chemistry, geology, mathematics, and physics; from the American Association for the Advancement of Science (AAAS); from the National Science Foundation (NSF); and from the developing institutions.

In 1970, a program of institutional grants for science in the black colleges was proposed. It was a formula-grant type of program and was to begin in 1972. The proposal came at a time when this particular type of support mechanism was on the decline. The foundation held meetings with representatives of black colleges to brief them about ongoing programs and to receive their suggestions and recommendations on the proposed programs.

The proposal to aid black colleges that was approved by the National Science Board and the administration was submitted to Congress as a comprehensive program of college-science improvement. This program was passed by Congress and was established in 1971–1972. It was the initial minority-targeted activity at the NSF and was implemented as the College Science Improvement Program-D (COSIP-D), for historically black four-year colleges and those institutions funded to serve other disadvantaged minorities. COSIP-D was budgeted at $5 million in the first year.

The primary purposes of COSIP-D were to accelerate the development of the undergraduate science capabilities in the historically black institutions and to enhance their capacity for self-renewal. The scope of the program was some 85 or so four-year institutions. The COSIP-D program was part of an existing college-science-improvement program restricted mainly to four-year colleges. In 1973, eligibility for participation in the COSIP-D program was expanded to include two-year institutions. Except for Dartmouth College and Pembroke College and institutions recently chartered by the Native American Indian nations, most institutions at that time that were founded to provide education for minorities were black. The eligibility criterion requiring that the institution be founded to provide education for disadvantaged ethnic minorities excluded most other institutions.

In 1974, additional expansion of the eligibility for participating in the COSIP-D programs led to the inclusion of those institutions that serve black, Spanish-speaking, American Indian, and other disadvantaged ethnic minorities. Currently, the scope of the program is of the order of 200 institutions enrolling some 400,000 students.

It is significant that the minority program at the NSF was aimed at an identifiable legal target: the institution. The institution, of course, served the target population.

The COSIP-D program has provided support for the following types of activities at minority institutions:

1. The testing of new educational procedures; the improvement of instructional facilities; the design of curricula, local courses, and curriculum improvements; the purchase of instructional equipment; and faculty development.
2. Financial support of student trainees.
3. Research initiation—financial support for scientific research.

Each institution has combined the above elements in its proposal to meet its objectives.

In addition to providing support at individual institutions, the COSIP-D program encourages cooperative activities, for example, broad projects coordinating national resources that enable faculty at minority institutions to make informed decisions regarding computer equipment, courseware, instructional techniques, and the use of computer-based instructional systems that may be employed in the improvement of science-education programs. As a result of such cooperative activities, Lincoln and Knoxville colleges have received funds to solicit expert assistance to design and develop audiovisual and other tutorial materials to aid in the teaching of science and mathematics on their campuses.

To date, in the four years of the COSIP-D program, $20 million has been spent on individual and cooperative activities; $4 million on research initiation; and $0.5 million on traineeships, for a total of approximately $24.5 million. The outcome in terms of the effect on numbers of students is yet to be assessed. The first cycle of three-year projects has only recently been completed.

Some of the questions that are addressed by the COSIP-D program are:

1. How can institutions be assisted in overcoming the effects of institutional isolation and a lack of adequate facilities and equipment?

2. How can institutions attract and hold promising scientists and scholars, an invaluable component of the strength of an institution?
3. How can institutions develop programs that establish effective instructional procedures for preparing students in science?

All of these questions comprise elements that contribute to the inadequate access that minority students have to careers in science.

Future program development for improving the access of minority students to science will require two steps (other than dollars):

1. Evolution of the universe of institutions.
2. Evolution of models of improvement.

Current estimates show that the major enrollment of minority students is outside of the NSF-targeted institutions. There is no evidence, however, that institutions with the greater percentage of minority students will produce the greater percentage of minority graduates.

19

Minority Programs Funded by the National Institute of Mental Health

MARY S. HARPER

In the discussion of the National Institute of Mental Health (NIMH) scientific activities in education and research, I shall present an overview of the status of research, training, and accomplishments of four minority groups,* and an overview of the NIMH and of the specific research activities of the Center for Minority Groups Mental Health Program (CMGMHP).

An analysis of the training of science and engineering professionals shows the following. For the period of 1930–1972, 244,829 doctorate degrees were awarded; of these, 14,663 (6%) were awarded to minority-group members. Of the 14,663 doctorates awarded to minority groups, 75% (10,987) were awarded to Orientals; 12.7% (1,860) to blacks; 9.6% (1,412) to Latins; and 0.7% (106) to American Indians.

Of the 14,663 minority-group members receiving doctorate degrees, approximately 28% were United States citizens. The trend of noncitizen minority groups' receiving the largest number of doctorate degrees has continued, as demonstrated in the 1973 report of doctorate degrees earned (National Academy of Sciences, 1974). In 1973, almost 4,000 of the 33,727 Ph.D.'s in the United States went to members of minority groups, but only 37% of these were United States citizens. The Orientals were the dominant group, followed by blacks, Latins, and American Indians.

When one tries to translate these figures into data on the United States labor force, it is difficult because the postgraduation plans of

*Minority groups serviced by the Center include: American Indians, Asian Americans, blacks, and Spanish.

MARY S. HARPER • Assistant Chief, Center for Minority Group Mental Health Programs, National Institutes of Mental Health, Rockville, Maryland.

many of the foreign citizens are uncertain. Estimates of the proportion of racial/ethnic groups added to the United States labor force from the 1973 Ph.D. graduates (United States citizens and immigrant-visa cases) are given in Table 1 (National Academy of Sciences, 1974, p. 11). Of the 4,000 doctoral recipients among the minority groups, only 8.8% were added to the United States labor force.

Currently, minority groups comprise 6.6% of the Ph.D.-level science and engineering labor force in the United States. Orientals comprise 5% of this group, blacks 0.8%, Latins 0.6%, American Indians less than 0.1%, and all other minority groups approximately 0.1%.

The employment categories for the new graduates are not proportionately distributed across the racial/ethnic groups. For instance, blacks, Latins, and American Indians tend to concentrate in educational institutions (63.7%, 55.3%, and 69.5%, respectively). Orientals concentrate in business and industry (28.3%).

Because of economic circumstances and an income that is generally 60% lower than that of nonminorities, it takes a minority citizen of the United States 11.3 years to get a doctorate (median time-lapse from baccalaureate to doctorate) (Bryant, 1970). The median time-lapse for noncitizen minorities is 8.4 years. Bryant found that 70% of the black Ph.D.'s required 10 years and more to get their degree (B.S. to Ph.D.). Of the black Ph.D.'s, 21% took more than 20 years to get their doctorate degree (time-lapse between B.S. and Ph.D.).

In the mental-health areas the increase in mental-health workers is given in Table 2. In spite of this progress, shortages still exist. These

Table 1. Estimated Racial/Ethnic Proportion of 1973 Doctoral Graduates Added to United States Labor Force

Racial/ethnic background	Number of cases	Percentage of total	Percentage distribution within minority group
White	27,226	91.2	—
Black	816	2.7	31.3
Oriental	1,387	4.6	52.8
Latin	251	0.8	9.6
American Indian	148	0.5	5.6
All other minorities	24	0.1	0.9
Minority total	2,629	8.8	100.0

Table 2. Increase of Those Employed in Mental Health, 1960 and 1975

Professional discipline	1960	1975
Social worker	26,000	60,000
Psychiatrists	14,000	25,000
Registered nurses	504,000	815,000
Psychologists	18,000	37,000

shortages will total some 40,000 professionals by 1981, even with the current rate of training (U.S. Senate, 1976).

There is a shortage of all mental-health workers, but the shortage among minorities is insurmountable: "Blacks represent 2 percent of the total 323,210 physicians, 1.8 percent of the 21,150 psychiatrists, 2.6 percent of the 22,501 doctoral psychologists, 5.6 percent of the 815,000 nurses, 10.9 percent of all social workers and 6.2 percent of the total 24,122 sociologists" (Brown and Ochberg, 1973).

There is a tremendous need to enhance the quality and quantity of minority professionals in mental health. For example, between 1920 and 1966, the 10 most prestigious departments of psychology in the United States granted 8 Ph.D.'s to blacks out of 3,767 Ph.D.'s awarded. Of these 10 departments, 6 did not grant a single Ph.D. to a black; yet most of the consultants on federal training and research programs came from these departments (Brown and Ochberg, 1973).

Faculty ranks in graduate departments are generally lower for minorities than they are for nonminority faculty of equal experience and education. For example, graduate departments in sociology in 1974 included 4.9% blacks, 0.2% American Indians, 1.0% Spanish-Americans, 2.2% Asian-Americans, and 18.5% women. Blacks tend to be in the lower ranks: 10.3% of the instructors, 1.2% of the lecturers, and 4.5% of the assistant professors. Asian-Americans have the highest rank: 2.3% of the associate professors, 2.7% of the assistant professors, 0.4% of the instructors, and 1.1% of the lecturers (Scientific Manpower Commission, 1975). In 1973, 13.4% of all minority doctoral scientists and engineers were engaged in research in educational institutions as their primary activity.

The federal government employs few Ph.D.'s from minority groups when compared to all employers jointly. This is certainly true for Orientals but only marginally true for blacks and Latins. In the 1973

"Comprehensive Roster of Doctoral Scientists and Engineers" (National Academy of Sciences, 1974), it was indicated that the federal government employs 18,431 doctoral scientists and engineers, of which 4% (785) are minorities. This 4% included 0.8% blacks, 2.8% Orientals, 0.5% Latins, and 0.01% American Indians.

In summary, the preparation and employment of minority workers in education, research, and government have increased but not at an even rate of growth for the American-born citizen, although the number of minority members in health care and in educational and correctional institutions has increased and in some instances doubled (Brown and Ochberg, 1973; U.S. Bureau of the Census, 1973a,b).

In 1971, the Secretary of the Department of Health, Education, and Welfare (HEW), Elliott Richardson, and Dr. Bertram Brown, Director of the National Institute of Mental Health (NIMH), announced the establishment of the Center for Minority Groups Mental Health Programs. The mental health of minority groups (American Indians, Asian-Americans, blacks, and Spanish) is now one of NIMH's priorities (Brown, 1970).

In the announcement of the establishment of the Center for Minority Groups Mental Health Programs, Secretary Richardson and Dr. Brown explained that the center would serve as a focal point for all activities within the institute that bear directly on meeting the mental-health needs of minority groups, including programs of research, training services, and demonstration projects (U.S. Department of Health, Education, and Welfare, 1970).

In the early planning phase, it was assumed that the center could provide leadership in the development and application of fresh knowledge from the view of the minority groups; lend support to the professional and nonprofessional training of minority-group members; stimulate support and evaluate mental-health service programs for minority groups; collect and disseminate information about mental-health problems and resources related to cultural, ethnic, and racial issues; and advocate priorities within the institute based on the mental-health needs of minority groups.

These missions are consistent with the general goals of the institute, responsive to needs forcefully outlined by minority-group components, and congruent with federal emphasis on targeted programs in areas of great need. The center supports mental-health projects in research, the development of mental-health professionals, and the utilization, evaluation, and delivery of mental-health services to minority groups.

The activities of the center include the processing, administration, and monitoring of grants and contracts; arrangements for intramural and extramural consultation; limited consultation to agencies, groups, and organizations on the mental health of minority groups; participation in the review, drafting, and editing of major mental-health policies and legislation; and the development and dissemination of information relevant to the mental health of minority groups.

In addition, the center supports the development, administration, and monitoring of five national research-and-development centers. These are the National Research and Development Mental Health Center for American Indians and Alaskan Natives, University of Oregon; the National Research and Development Mental Health Center for Asian Americans, University of California at San Diego; the National Research and Development Mental Health Center for Blacks, Howard University; Fanon National Research and Development Mental Health Center for Blacks, Drew Postgraduate Medical School, Los Angeles; and the National Research and Development Mental Health Center for Hispanics, University of California at Los Angeles.

The principal projects of the center that involve the preparation of minorities in education and research include the preparation and supervision of minority researchers in the five national research-and-development mental-health centers; the development of curriculum content and pilot programs in pediatric psychology; graduate training programs in social work and psychology with educational modules for mental-health programs in inner-city and urban settings for minority groups; training programs for paraprofessionals and mental-health aides to work with indigenous groups; and the administration and monitoring of the five national fellowship programs for minority groups.

The five national fellowship programs in mental health and social and behavioral science were conceived and developed to enhance the quality and quantity of mental-health workers in research for minority groups. A review of the literature indicates that most of the research relating to minority groups has been of a deficit/pathological or "blaming" nature (Caplan and Nelson, 1973; Ryan, 1971). Most of the persons conducting research on minority groups are not minority researchers. In order to ameliorate this condition, the CMGMHP supported a grant to five of the national professional associations: the American Nurses Association, Kansas City, Missouri; the American Sociological Association, Washington, D.C.; The National Council of Social Work Education, New York; the American Psychological Association, Washington, D.C.; and the American Psychiatric Association, Washington, D.C.

Each of the fellowship programs is funded for a five-year period, and each has a director and a clerical support staff. The fellowship programs have a 10-person to 15-person advisory committee of minority behavioral scientists competent in research training and administration and consultation in research. These committee members are also knowledgeable about the mental-health needs of minority groups and have previous experience in an academic institution in graduate education.

The committee selects a minimum of 10 fellows annually. Each fellow is awarded tuition and fees, a dependent allowance, fees for books, etc. The total amount of each stipend does not exceed $7,500. The student selects the academic institution or the supervised learning setting. However, the student must be accepted into a graduate research program prior to the submission of an application to the fellowship program. The student's major area of concentration must be in mental health–social-behavioral science–psychiatric research involving minority groups or areas of minority groups' concerns or needs. It is anticipated that persons admitted to the fellowship programs will be able to complete their doctorate degree in three to four years. Upon completion of the program, it is anticipated that each fellow will become actively engaged in conducting, teaching, or consulting in research involving minority groups.

The program of the American Sociological Association (ASA) is cofunded by the Office of Educational Equity of the National Institute of Education (NIE). The fellows supported by the NIE are persons who are majoring in the sociology of education. The fellows for the sociology of education component are selected and monitored by the National Advisory Committee.

The fellowship programs are in various stages of development. One program is in its fourth year and the others are in their second and third years. The fellows of each discipline meet at the national convention of their association to discuss their research proposals and/or findings. Two of the fellowship programs have experienced excellent collaboration with some of the universities, which have provided one or two minority students with tuition waivers. Two of the associations have matched some of the federal funds, thereby allowing an increase in the number of fellowships to be awarded. One professional association invested the funds from a foundation in the fellowship program. With the support of the universities, associations, and foundations, three of the fellowship programs have been able to award almost twice the number of fellows supported by the NIMH. It is now projected that about 120 fellows will be prepared through the fellowship program described here. Only 50 will be supported by NIMH funds.

In summary, the CMGMHP supports training in research through individual grants, through the five research-and-development mental-health centers, and through the five national mental-health fellowship programs. There is such a need for an improved quality and quantity of research for minority groups that the activities of the center in this area should be doubled.

REFERENCES

Brown, B. S. The people and the priorities. Paper presented at the annual meeting of the National Association of Mental Health, Los Angeles, California, November 1970.

Brown, B., and Ochberg, F. Key issues in developing a National Minority Mental Health Program at NIMH. In C. V. Willie *et al.* (Eds.), *Racism and mental health*. Pittsburgh: University of Pittsburgh Press, 1973. Pp. 555–573.

Bryant, J. W. A survey of black American doctorates. Ford Foundation Special Project in Education, 1970.

Caplan, N., and Nelson, S. D. On being useful—the nature and consequence of psychological research on social problems. *American Psychologist*, 1973, March 1973, pp. 199–211.

National Academy of Sciences, Commission on Human Resources of the National Research Council. Minority groups among United States doctorate level scientists, engineers and scholars, 1973. Washington, D.C., December 1974.

Ryan, W. *Blaming the victim*. New York: Pantheon, 1971.

Scientific Manpower Commission. Professional manpower: A resource service of data on women and minorities. Scientific Manpower Commission, 1776 Massachusetts Avenue, NW, Washington, D.C. 20036, 1975.

U.S. Bureau of the Census. U.S. census of population 1970. DC(A)-4E, 1973a, p. 36.

U.S. Bureau of the Census. U.S. census of population 1970: Persons in institutions and other group quarters, subject reports. PC(2), Table 3, 1973b.

U.S. Department of Health, Education, and Welfare. Press release, November 19, 1970.

U.S. Senate. Excerpts from Senate Report on Labor—HEW Appropriation Bill for FY76, 1976.

20

The Importance and Impact of Funding Science Programs at Minority Institutions

MILES MARK FISHER, IV

The past few years have resulted in the raising to national consciousness of the need to encourage more blacks and other minorities to pursue science-oriented careers. With this thrust has come a recognition of the importance and impact of funding science programs at minority institutions. Minorities in general have not fared well in their participation in the various science-oriented disciplines in American higher education. Blacks in particular have indeed been limited in their participation.

This statement will deal with the importance as well as the impact of funding science programs at minority institutions. Emphasis is on the historically black colleges and universities. Let us look at this particular subset of minority institutions.

There are some 107 historically black colleges and universities of this nation, representing private two-year and four-year institutions and public two-year and four-year institutions, as well as graduate and professional schools, located in 15 southern states, four northern states, and the District of Columbia. These institutions enroll more than 200,000 students and graduate more than 30,000 students annually with undergraduate, graduate, and professional degrees. Since 1966, these institutions have awarded more than a quarter of a million undergraduate, graduate, and professional degrees. They have been the providers of equal educational opportunity to thousands of their students.

Higher education has been the means by which students from low-

MILES MARK FISHER, IV • Executive Secretary, National Association for Equal Opportunity in Higher Education, Washington, D.C.

income backgrounds have been able to attain upward mobility in society. The black colleges represent an existing mechanism that can be used to intensify efforts to equalize opportunity for all students. As late as 1974, black colleges graduated over 50% of all black Americans earning baccalaureate degrees. As late as 1972–1973, 60% of all blacks who earned doctorates had previously received their baccalaureate degrees at black colleges and universities. There are only estimates for baccalaureate degrees awarded outside the black colleges, as no comprehensive survey of black graduates outside of the black colleges has been undertaken.

In spite of this achievement and productivity, there has been a historical neglect of these institutions across the years, a neglect that must be addressed and rectified at this time. Throughout the advent of *Sputnik* and the mass infusion of funds into institutions of higher learning, minority institutions have not been included in the real programmatic support. Majority institutions, on the other hand, have been able to develop and expand their capabilities for scientific education and research. Unfortunately, opportunities for minority students to pursue graduate work at majority institutions have been minimal. In addition, minority students require financial assistance, which is frequently not available.

In a study (Institute for Services to Education, 1975) of the fall term for the academic year 1975–1976, the total enrollment in black colleges increased 11.8% from the fall enrollment in the academic year 1974–1975, increasing from 189,001 to 211,366 students. The total freshmen enrollment rose from 46,706 students in 1974 to 54,221 in 1975, an increase of 16.1%. The pattern of growth observed over the past nine years has increased significantly during the 1975–1976 academic year.

When the national enrollment from fall 1976 to fall 1975 is considered, the black colleges increased by 11.8% as compared to 8.9% for all colleges nationally. The 16.1% increase in freshman enrollment was nearly two and a half times the 6.5% increase in colleges nationally. These data indicate that black colleges continue to be increasingly attractive to minority students.

In an American Council on Education survey (Astin, King, and Richardson, 1975) of the 1975 fall freshman classes at black colleges, 40% of the students expressed an interest in the science majors shown in Table 1. In the area of probable career occupation, it was reported that 23% of the students indicated career choices related to science (Table 2).

Table 1. Majors Chosen by 40% of Fall 1975 Freshman Classes at Black Colleges

Field	Percentage
Biological sciences	3.7
Engineering	4.6
Health professions	7.4
History, political science	4.3
Mathematics and statistics	0.5
Physical sciences	1.6
Social sciences	11.2
Other technical fields	6.9
Total	40.2

Students in black colleges appear inclined to pursue major fields of study requiring strong science backgrounds as well as career occupations that demand such an academic base.

In spite of this continuing tendency of blacks to aspire to fields of study and career occupations that require solid scientific backgrounds, there is an inadequate representation of blacks and other minorities in the sciences. The deficit is reflected at the undergraduate level, the graduate level, and the professional level.

Graduate enrollments indicate that blacks and other minorities, excluding Asian-Americans, comprise a small percentage of the science and engineering pools. At a time when policy makers suggest that there is an oversupply of Ph.D.'s in the sciences, they have not taken into consideration the underrepresentation of blacks and minorities in these areas. Trends in the major society do not necessarily reflect the trends and needs of the minorities. The specific needs of minority groups should be kept in mind when manpower projections are made for the sciences.

Table 2. Careers Chosen by 40% of Fall 1975 Freshman Classes at Black Colleges

Field	Percentage
Doctor (M.D. or D.D.S.)	4.2
Engineer	3.7
Health professionals (non-M.D.)	10.2
Nurse	4.5
Research scientist	1.0
Total	23.6

Shortages in human resources in the professional areas are quite severe. In the academic year 1975–1976, there was a decline in the number of first-year medical students. If this trend is not reversed, there will be a precipitous decline in the number of physicians needed to render service to large numbers of our population.

In the past 100 years, 80% of the nation's black physicians and dentists have been trained at Meharry Medical College and the Howard University School of Medicine. Some additional statistics should be noted:

A. 2% of all physicians are black.
B. 2.5% of all dentists are black.
C. The majority of black pharmacists are being trained at Xavier University, Texas Southern University, Florida A.&M. University, and Howard University.
D. 85% of all black veterinarians in this country have been trained at Tuskegee Institute.

The federal government has taken several initiatives to provide funding for science programs at minority institutions. Some of these programs are reviewed in this section (see the papers by Zora J. Griffo, James W. Mayo, and Mary S. Harper).

Science and technology are becoming increasingly important as our nation begins its third century. Minorities are not adequately represented in science and technology. If the fundamental premise of equal opportunity for all citizens is indeed to become a reality, it is the responsibility of the government to ensure that such opportunities are available to the economically and socially disadvantaged of our nation.

REFERENCES

Astin, A. W., King, M. R., and Richardson, G. T. "The American freshman," national norms for fall, 1975. Cooperative Institution Research Program, American Council on Education and the University of California at Los Angeles, 1975, p. 44.

Institute for Services to Education, Inc. Management Information System News, Research Profile Preliminary. Fall 1975. Report on enrollment in historically black colleges, 1975, p. 2.

V. Affirmative Action: Myth or Reality?

21

Affirmative Action: A Congressional Perspective

EDWARD R. ROYBAL

The essence of good government lies in its humanity—in its commitment to help the oppressed and the poor find economic well-being and equality. Affirmative action is a manifestation of that commitment. Its justification derives from the history of exclusion and discrimination suffered by millions of Americans who have sought, and been denied, equal access to employment and educational opportunity in this country. Affirmative action seeks to correct past and present discriminatory patterns and their effects. The concept dates back to the Wagner Act, which 40 years ago required "affirmative action" against employers whose antiunion activities violated the law. Title VII of the Civil Rights Act of 1964 has applied this concept to discrimination in employment based on race, national origin, and sex. Executive Order 11246, issued in 1965, barred employment discrimination in the federal government and by federal contractors and subcontractors.

In addition, Title VI of the Civil Rights Act prohibited racial and national-origin discrimination in the participation and enjoyment of benefits in federally assisted programs. In 1971, the Public Health Service Act barred sex discrimination in admissions to health-related training activities and schools of nursing. In 1972, Title IX of the Education Amendments prohibited sex discrimination in most education programs.

The foundation for all of these laws emanates from the spirit and the language of the Constitution. The Fourteenth Amendment, for example, guarantees "equal protection of the laws" and is designed to end the oppression of inequality and class privilege.

EDWARD R. ROYBAL • Member, U.S. House of Representatives, Democrat—California.

Under the aegis of the Fourteenth Amendment, affirmative action goes far beyond the simple condemnation of discriminatory and unequal policies. It is not enough for our institutions to advertise as "equal-opportunity employers" or promise an end to discrimination in hiring. By themselves, these expressions are but paper assurances.

If "equal protection of the laws" is to have meaning, it must include affirmative steps to reverse the status quo—to overcome the exclusion of minorities and women from the full enjoyment of health and medical opportunities.

Affirmative action also calls for a deep and lasting commitment on the part of our educational leaders to provide academic opportunity in all areas. But our institutions have yet to move in that direction. In fact, very little if any civil-rights enforcement has been occurring on our campuses.

1. In a 1975 report, the General Accounting Office (GAO) concluded that the Department of Health, Education, and Welfare "has made minimal progress in making sure that colleges and universities have acceptable affirmative action programs" (U.S. General Accounting Office, 1975, p. 1). It cited HEW's failure to send "show-cause" notices to noncomplying institutions; failure to conduct preaward reviews; overly prolonged negotiations; and failure to adequately enforce the program at the Berkeley campus (U.S. General Accounting Office, 1975, Chaps. 2 and 3).

2. In 1975, the House Subcommittee on Equal Opportunities held extensive hearings on affirmative action at higher educational institutions. The subcommittee found "no evidence to indicate that academic institutions are so unique as to warrant equal employment regulations different from other federal contractors and insisted that there be no 'special cases' in the enforcement of the Executive order" (Hawkins, 1976). The subcommittee noted also that the enforcement of equal opportunity at institutions of higher education has been ineffective and that federal-contract compliance has been deficient.

3. The U.S. Commission on Civil Rights (1975, p. 367) found, that same year, that "the inadequacy of HEW's enforcement effort . . . permits the continuation of practices which result in the denial of equal education and employment opportunities to women and minorities."

The commission noted that HEW had failed to carry out "in depth and regular" compliance reviews of higher institutions receiving federal funds. Over a 10-year period, HEW had reviewed less than 30% of all campuses covered under Title VI. Further, the commission found that during fiscal year 1974, HEW conducted compliance reviews of only 6% of the campuses covered under the Executive Order (U.S. Commission on Civil Rights, 1975, pp. 368–370). Like the GAO, the commission expressed concern over HEW's consistent failure to issue show-cause notices to noncomplying institutions. For instance, from 1971 through 1974, HEW issued only two notices despite the uncovering of numerous violations (pp. 370–371). The commission was critical of HEW's willingness to accept only the assurances of institutions to develop an affirmative-action plan, although the standard of compliance has always been the actual existence of the plan (p. 371). HEW's retreat displayed a kind of convoluted logic involving a plan to have a plan, which was given classic expression in the belabored Berkeley agreement.

4. There has been very little evidence of a federal or an institutional commitment to equal opportunity in the health area. Most educational institutions lack any formal and effective mechanism to ensure that federally funded training and research activities on campus actively involve minorities and women.

The National Institutes of Health spend well over $2 billion for health research and training; $136 million will go directly to training activities this year alone. Despite this sizable infusion of federal funds each year, the NIH has yet to conduct preaward checks or to monitor grantee institutions to determine the adequacy of their equal-opportunity efforts.

The National Cancer Institute controls 15% of the NIH's training budget; and yet it conducts no monitoring, only "public relations," which it has assigned to its equal-employment office.

The same pattern of indifference exists among other health-related agencies. The Office of Human Development spends $$22 million a year for the training of 9,000 rehabilitation workers in such areas as mental illness, medicine, nursing, counseling, and speech pathology and audiology. And yet agency officials admit that they conduct no preaward checks.

What we clearly have is an absence of substantial opportunity for minorities and women to enter the sciences and the medical professions.

1. An analysis of data furnished by the National Academy of Sciences (1975) has shown that minorities are severely underrepre-

sented in federally funded graduate training programs. Although minorities comprise nearly 20% of the total population, they represent only 5% of graduate-degree recipients. (Black Americans constitute 2.7%; Mexican-Americans and Puerto Ricans, 0.8%; American Indians, 0.6%; and Asian-Americans, 0.9%.) Further, the data reveal only a slight increase in science and engineering Ph.D.'s for minority citizens. Over a 40-year period, minority representation in this area has moved from an imperceptible 0.6% for blacks and Latinos to a marginal 1.4% (National Academy of Sciences, 1974).

2. Despite the fact that the federal government provides nearly $1 billion in medical financial support, minority enrollment in medical schools remains deficient. For the academic year 1974–1975, black and Latino enrollment in medical schools represented 7.8%, with 1.5% Latino. Further, black and Latino first-year enrollments showed a decline from 9.5% to 8.8% for 1975–1976 (Association of American Medical Colleges, 1975).

3. And black and Latino enrollment in public-health schools comprised only 7% of total enrollment for 1974–1975 (Association of Schools of Public Health, 1975).

4. In the field of physician's assistants, the level of black and Latino graduates is expected to decline from 7.3% to 4.6% in 1976 (Fisher, 1975).

5. In the employment area, we find a similar pattern of underrepresentation. Equal-employment data from 1974 for nearly 4,200 employers in the area of health services showed that blacks and Latinos were severely underrepresented in the health professions. Blacks comprised only 4.4% and Latinos 1.8% of health-professional staff (Equal Employment Opportunity Commission, 1974). While their employment picture improves for white-collar and office-manager positions, they remain relegated to the lower occupational categories.

Similarly, figures on federal employment from 1974 showed that blacks and Latinos represented only 5.4% of the 8,300 medical and dental officers employed governmentwide (U.S. Civil Service Commission, 1974). Further, Latinos comprised only 1.4% of the Health Services staff and less than 1% in Health Resources and in Alcohol, Drug Abuse, and Mental Health. Further, Latinos represented 0.6% in the entire National Institutes of Health (U.S. Department of Health, Education, and Welfare, 1975).

It has been argued that facts and figures do not justify changes in admission standards and the pursuit of academic excellence, that these ideals must be preserved at all cost. But such an argument suffers from some basic flaws. First, the argument implies that changes in standards are detrimental. This position reflects a *status quo* mentality—one that lacks critical self-analysis, a prerequisite of a scientific approach. What are these standards? They are highly subjective impressions that lack validation and are based on judgments of who shows the greatest intellectual or creative promise or who would be the most effective health practitioner. But the danger is that such standards of merit are likely to reflect the particular biases of the white-dominant educational elite that selects the students and faculty.

The argument also ignores a basic educational ideal to which we have already committed billions of federal dollars—and that is the full development of the scientific and health resources among our entire population. That includes the active recruitment of talented and promising minority individuals. If it means the establishment of special recruitment and outreach efforts, or of extended educational and entry programs, then it must be done and can be without the jeopardizing of standards. For the alternative would be a return to white-only medicine and the inestimable loss of talent, resources, and excellence of those with minority backgrounds.

The argument overlooks the recent findings presented in the *Journal of the American Medical Association* (1975). The *Journal* reported that the retention rate among minority students in medical schools was "at a high level of most groups." For minority students enrolled in medical schools in the past three years, the retention rate was 94%. A 1975 study of Chicano medical students showed that 85% came from the top third of their high-school classes, which compared favorably with figures on nonminority students (Kavfert, Martinez, and Quesada, 1975).

The point that needs to be hammered over and over again is: How can we continue to justify such a tremendous loss of talent and potential by ignoring the 20% minority and 50% female population in this country? How can we continue to deny equal access and opportunity to a large segment of our taxpaying population who support the administrators, the researchers, the health practitioners, and the teachers at our educational and medical institutions?

The fact is that minority students are doubly faced with the economic demands of obtaining an education, especially in the medical professions. Minority families do not have the financial resources to help cover tuition and other related expenses. The data show that these families represented only 3.5% of the total reported family contributions to student assistance during 1973–1974 (National Academy of Sciences, 1975).

In a study of midwestern colleges, one researcher found that 80% of minority students needed financial support and that over three times as many minority freshmen, as compared to other freshmen, needed full financial assistance (Willie and McCord, 1972; College Entrance Examination Board, 1970).

A recent sampling of graduates of Los Angeles city high schools revealed that only 60% of low-income-area students who were eligible to enroll at the University of California or similar institutions were able to do so; 27% went to work. By contrast, 80% of eligible students from high-income areas entered four-year colleges; only 9% went to work (McCurdy, 1975).

It is very likely that this negative economic pattern will continue, especially during periods of high unemployment, which have a disproportionate effect on minority families.

As devastating as this situation may be, President Ford's budget for fiscal year 1976 proposed to add insult to injury by cutting student assistance in the health professions by 50%. The President's budget proposed to:

1. Abolish supplemental-opportunity grants and direct loans for students.
2. Reduce work–study funds by 50%, while raising the institutional matching share from 20% to 50%.
3. Cut back on special programs for the disadvantaged and eliminate university community services.

In addition, the President's budget sought a 37% reduction in health services to our communities. In line for cuts were programs for maternal and child health, emergency medical services, and community health centers. The executive budget called for a 36% slash in community mental-health and alcoholism services and completely eliminated health grants to the states. These proposals represented a drastic retreat in student assistance and health services for the disadvantaged. They spelled disaster for those students and families already crippled by lack of economic and health resources. As long as health and other living expenses skyrocket and student financial assistance dwindles, the level of minority enrollment and graduation, particularly in the health professions, will decline.

This is, indeed, a pessimistic reflection on the current state of affirmative action and equal opportunity in this country. It is for this reason that our colleges and universities must be firmly committed to the active involvement of minorities and women at all institutional levels, including the positions of leadership and decision making.

I recommend that each educational and health institution be willing to:

1. Conduct a self-analysis of the minority and female composition of its student body, its faculty, and its administration in order to identify areas of deficiency and their causes.
2. Evaluate admission standards and employment prerequisites, such as degree attainment, publication achievements, and tenure.
3. Develop goals and timetables for overcoming underrepresentation.
4. Develop a plan of affirmative action in the area of educational services, financial assistance, and research and training opportunities.
5. Provide for the involvement of minorities and women in the development and the assessment of all affirmative goals and studies.
6. In the case of publicly supported systems of higher education, develop a statewide plan of educational opportunity for all students.
7. Develop year-long cooperative educational programs with local school districts to encourage and train minorities and women to enter the health and scientific fields.
8. Develop close working arrangements with community organizations in fostering recruitment and outreach programs for disadvantaged students.
9. Require an annual self-evaluation to determine areas of success and failure in meeting affirmative-action and equal-opportunity objectives.

It is clear that these recommendations, if fully realized, would cause changes in institutional leadership and decision making. This is, to a great extent, the reason for the current reluctance about and resistance to affirmative action. And yet the pursuit of excellence requires us to bring new perspectives and ideas into our institutions of higher education. The pursuit of excellence demands that we actively recruit and seek out new talent and creative minds from the untapped and disadvantaged communities of this nation. This is the significance and necessity of affirmative action.

REFERENCES

Association of American Medical Colleges, Division of Student Studies. U.S. medical school enrollments, 1971–72 through 1975–76. October 1976.

Association of Schools of Public Health, Data Collection Center. U.S. students by race and program of study. Table B12 of U.S. schools of public health, 1974–1975. October 1975.

College Entrance Examination Board. Midwestern higher education surveys, Report M-1, 1970.

Equal Employment Opportunity Commission. 1974 EEO-1 report summary nationwide industries—health services, 1974.

Fisher, D. W. (Executive Director, Association of Schools of Public Health). Age, sex, ethnic origin as a function of year of graduation. Table I of a paper presented at the Conference on Current Information on Health Manpower, Tarrytown, New York. April 16–18, 1975.

Hawkins, A. F. Hawkins attacks Higher Education Committee (news release). February 4, 1976.

Journal of the American Medical Association. 75th annual report—medical education in the United States, 1974–1975. *Journal of the American Medical Association, 234*(13), 1339, and Table 14: Students admitted 1972–73 through 1974–75 and still in medical school or graduated. June 1975.

Kavfert, J., Martinez, C., Jr., and Quesada, G. A preliminary study of Mexican-American medical students. *Journal of Medical Education,* 1975, *50,* 859–860.

McCurdy, J. Many of poor eligible for UC fail to enroll. *Los Angeles Times,* November 12, 1975, CC, Part II.

National Academy of Sciences. Native-born U.S. citizens in the comprehensive roster, showing minority racial/ethnic groups. Table 7 of minority groups among United States doctorate-level scientists, engineers, and scholars, 1973. Washington, D.C., December 1974, p. 22.

National Academy of Sciences. Statistical profile of doctoral recipients, by racial or ethnic group and U.S. citizenship status. Table 5 of summary report 1974: Doctoral recipients from United States universities. Washington, D.C., June 1975, pp. 22–23.

U.S. Civil Service Commission, Bureau of Manpower Information Systems. Employment data as of November 30, 1974, compiled as requested by author. Received February 4, 1976.

U.S. Commission on Civil Rights. The Federal Civil Rights Enforcement Effort—1974, Volume III: To ensure equal educational opportunity. January 1975.

U.S. Department of Health, Education, and Welfare. Employment data as of December 1975 compiled as requested by author. Received February 13, 1976.

U.S. General Accounting Office. More assurances needed that colleges and universities with government contracts provide equal employment opportunity. August 25, 1975.

Willie, C. V., and McCord, A. S. *Black students at white colleges.* New York: Praeger Publishers, 1972.

22

A National Policy for Affirmative Action in Higher Education

MARGARET S. GORDON

Affirmative action in higher education has been a focus of increasingly serious controversy in recent years. Part of the trouble lies within institutions of higher education themselves, where there is frequently a conflict between women and minorities, on the one hand, pressing for more vigorous enforcement of affirmative-action policies and, on the other, vigorous opposition by some (but by no means all) white male faculty members to the enforcement of affirmative-action plans. Part of the trouble—and probably the most serious part—stems from the cumbersome and inept federal enforcement of affirmative action in relation to higher education.

Much of the controversy centers on the concept of goals and timetables. Critics of affirmative action policies do not object to more vigorous attempts to pursue nondiscriminatory procedures in the appointment and promotion of faculty members, but they argue that the goals and timetables that set forth plans for adding women and members of minority groups to the faculty are indistinguishable from quotas. Supporters of goals and timetables, on the other hand, point out that preferential hiring of women and minorities is not required by federal regulations—in fact, it is proscribed—and that institutions are not subject to federal penalties if goals and timetables have not been achieved, provided they can show that "good-faith" efforts have been made to achieve them.

The controversy over federal enforcement reached a critical point in June 1975, when the Office for Civil Rights (OCR) in the U.S. Depart-

MARGARET S. GORDON • Associate Director, Carnegie Council on Higher Education, Berkeley, California.

ment of Health, Education, and Welfare—the agency responsible for the enforcement of affirmative action in higher education—warned 29 universities that it might withhold their pending federal contracts unless they quickly produced acceptable affirmative-action plans or agreed to follow a model approved by the agency. The crisis was resolved through a compromise that permitted most of the institutions to sign a brief agreement that provided for further meetings between representatives of the institutions and the OCR, resulting in the submission of revised affirmative-action plans 30 days after the meetings.

The objections of the universities to the OCR's demands in June 1975 centered on (1) the requirement of an enormous mass of statistical data, some of it unavailable because it related to past recruitment and appointment patterns, department by department, and (2) the resemblance of the model plan that the universities were asked to develop to the plan that had been approved for the University of California's Berkeley campus. The Berkeley plan had been the subject of much criticism because, on the basis of a comparison of availability ratios and utilization ratios for each of the 75 departments on the Berkeley campus, the plan set goals for adding women in only 31 departments, blacks in 1 department, Asians in 2 departments, and Chicanos and American Indians in no departments. These extremely limited results for minority groups resulted because members of individual minority groups were so meagerly represented in so-called availability pools in individual disciplines that the analysis showed a need for the addition of only a small fraction of a person in most instances (Administrative Conference of the United States, 1975).

My comments are based in large part on the Carnegie Council's report entitled *Making Affirmative Action Work in Higher Education,* which was made public in August 1975 and appeared in printed form in October (Carnegie Council on Policy Studies in Higher Education, 1975). We began work on that report early in 1975, and, among other things, conducted a survey of affirmative-action policies in a representative sample of institutions of higher education, receiving responses from approximately 132 colleges and universities. The results indicated that the great majority of the respondent institutions had developed affirmative-action plans and that most of the larger institutions had submitted their plans to the OCR for approval. However, only a small percentage of the plans that had been submitted had been approved, and the institutions almost universally reported lengthy delays—sometimes running to several years—between the submission of their plans and final approval. Significantly, also, many of the plans did not specify goals and timetables on the narrow departmental basis that was used in the development of the Berkeley plan, especially for minority groups, and a number of large universities with comprehensive plans expected to add relatively more members of minority groups than did Berkeley.

The council's report presented two sets of recommendations, the first addressed to institutions of higher education and the second to the federal government. We stressed the need for institutions to assume the initiative in developing and enforcing affirmative-action plans, pointing out that there had clearly been a past record of indifference to the aspirations of women and of minority groups to faculty status on campuses. Our report set forth the elements of a "good" affirmative-action plan, stressing the point that the most important requirement was a set of procedures that would ensure the involvement of decision makers at all levels of the institution in furtherance of the objectives—the board of trustees, the administration, the academic senate, faculty members in individual schools and departments, and all others involved in the selection and promotion of academic and nonacademic employees. In my observation, the dedication of the president or chief campus officer to the objectives of affirmative action makes all the difference in whether or not the campus develops a good plan and makes significant progress in achieving its objectives.

The Carnegie Council's report emphasized the point that a major obstacle to the achievement of the goals of affirmative action, especially in the case of minority groups, was to be found on the supply side, rather than on the demand side, of the academic job market. We therefore stressed the need for vigorous efforts to increase the supply of qualified individuals for academic employment, through the active recruitment of women and minority-group students by graduate and professional schools and programs aimed at providing them with financial aid and special assistance in overcoming any defects in their preparation for graduate study.

This emphasis on the need for increasing the supply of qualified persons for faculty employment has been widely accepted as essential in the case of minority groups but has been challenged as a "cop-out" by some spokesmen for women. There is general recognition of the fact that the proportion of blacks, Chicanos, and American Indians among Ph.D. recipients has been so small historically that it will be many years before they will be present on university faculties in anything like their proportions in the general population. In fact, there has been intense competition among leading universities in the hiring of the relatively few qualified minority faculty members. And because a substantial proportion of minority-group aspirants to graduate and professional degrees have low incomes and sometimes have had inferior undergraduate preparation, their need for special encouragement and assistance has not been disputed.

On the other hand, some spokesmen for women have argued that there is an ample supply of qualified women in a number of academic disciplines, that the unemployment rate among women with doctor's degrees has been higher than among men in recent years, and that the main problem lies in the continuation of discrimination on the demand side of the academic job market. There is little question that this objection is valid for some fields, such as English literature and history, where there are substantial numbers of qualified women who have been badly hit, like some of their male colleagues, by the adverse job market that has prevailed for the last six years in those fields in which there is practically no demand for holders of doctor's degrees outside of academic institutions, that is, in government and industry.

However, when the Carnegie Council report referred to the "supply gap," it was using the concept of inadequate supply in a special sense, that is, the difference between the percentage of women and minorities among recent Ph.D. recipients (about 18% in the case of women, perhaps 4–5% in the case of minorities) and their proportions in the labor force (38% and 14%, respectively) (National Research Council, 1974a,b; U.S. President, 1974). Even so, we probably should have made it clearer than we did that there *is* an ample supply of women in the humanities and in some of the social sciences, whereas in the natural sciences and in such fields as economics and engineering, women continue to make up a small percentage of Ph.D. recipients, despite some notable progress in the last few years.

Moreover, those who have implied that the aim of the Carnegie Council's report was to shift attention away from the demand side and to concentrate it on the supply side have clearly misread the report. The report calls for the vigorous pursuit of goals and timetables until such time as the existing wide disparities in the proportions of women and minorities among faculty members in various fields and types of institutions are greatly diminished. Nor did we intend, as some critics have implied, that federal-contract penalties should be extended to apply to the deficient representation of women and minorities among graduate and professional students. Rather, we had in mind that there should be some process of providing special credit or recognition to universities and colleges that developed programs especially designed to encourage the graduate and professional education of women and minorities. In this connection, it should be remembered that discrimination on the basis of sex in graduate and professional schools is prohibited by Title IX of the Education Amendments of 1972, while discrimination on the basis of race is prohibited by Title VI of the Civil Rights Act of 1964.

While endorsing the concept of goals and timetables, the council recommended certain changes in the manner in which they were formulated. Its report called for anlaysis to be undertaken for each academic department but stated that "decisions as to whether goals and timetables should be formulated for individual departments, or, as would usually be more appropriate, for groups of related departments or schools, or for the entire campus, should be made only after careful study of the probable overall results for all units and for the campus as a whole." I resisted this type of recommendation at first, because decisions to hire or promote are initiated by departments, and I did not favor a change that would take departments "off the hook." But after analyzing the results of the Berkeley plan, I became convinced that its department-by-department approach did in fact take nearly all departments on the Berkeley campus off the hook in the case of minority groups. Furthermore, our recommendations contemplated that each department's performance in hiring and promoting women and minorities would be scrutinized by affirmative-action administrators and campus committees, even though goals and timetables were formulated for groups of departments.

In fact, Department of Labor regulations have been altered to permit the combining of "departments having similar disciplines" in a revised "Format for Development of an Affirmative Action Plan by Institutions of Higher Education," which was transmitted by the Office of Federal Contract Compliance in the Department of Labor to the Office for Civil Rights in HEW on August 25, 1975.

I would stress the point, however, that we did not recommend that departments should *invariably* be combined. There is a strong case for departmental goals and timetables for women in certain fields, such as English and history, in which women receive a substantial proportion of the doctor's degrees.

In discussing other aspects of federal policies, the Council's report emphasized, in addition to the manner in which goals and timetables were determined: (1) the inadequacy of staff in the federal agencies most directly concerned, and especially in regional offices; (2) the inexcusable delays and backlogs of cases; (3) the proliferation of federal agencies involved; and (4) the fact that the only penalties available under Executive Order 11246—the withholding of contracts and debarment from future contracts—were so severe that they had never actually been used.

One of our most important recommendations was for the development of a more flexible set of sanctions for noncompliance with affirmative-action enforcement requirements in higher education. The council believes that the provisions for the withholding of federal contracts should be retained for use in cases in which more limited sanctions have failed to have an effect but that there should be milder, graduated sanctions related to the seriousness of the offense that could be imposed before this more severe sanction is used. This recommendation has also been made by the Administrative Conference of the United States and by a number of individuals (Vetter, 1974). There should also be an opportunity for a hearing before any sanction is imposed.

Second, we made a group of interrelated recommendations aimed at reducing the number of federal agencies involved and the overlapping and duplication of regulations, including requirements for reporting detailed data to a number of different agencies. These recommendations included: (1) giving the Secretary of HEW final authority to approve affirmative action plans and to impose sanctions on institutions, thereby avoiding the additional delays and uncertainties involved in the power of final review of certain important decisions by the Secretary of Labor; (2) the concentration of authority for processing complaints, except in equal-pay cases, with the Equal Employment Opportunity Commission and amendment of Title VII to give that agency the power to issue cease-and-desist orders; and (3) recommendations for coordinated guidelines and data-reporting requirements and for a more decisive role of the Equal Employment Opportunity Coordinating Council in coordinating federal action and eliminating the processing of a complaint by several agencies simultaneously.

Next, let me emphasize the need for development within these agencies of a staff that is knowledgeable about higher education, along with the recommendation that within the OCR, the staff members concerned with higher education should largely be centered in Washington and that important decisions, especially those involving the imposition of sanctions, should be made in the head office.

Some may ask why we did not recommend the concentration of all matters relating to nondiscrimination and affirmative action in higher education in a single agency. The answer is that the complex history of legislation and of the Executive Order makes it extremely unlikely that this could be accomplished, or perhaps even doubtful that it *should* be accomplished. Moreover, there would remain rights of bringing complaints before state agencies and both state and federal courts, which we were not inclined to question.

Finally, let me mention our recommendation that the coordinated federal regulations should include a requirement that all institutions of higher education develop internal grievance procedures for all employees, as is now required under Title IX regulations. Carefully framed grievance procedures that observe accepted standards of due process and that are available to all employees are now widely regarded as a desirable personnel practice.

We were much more cautious on the issue of whether federal agencies should *require* complainants to exhaust internal grievance procedures before filing a federal complaint, recommending instead that federal agencies should *encourage* complainants to exhaust internal remedies, where they were deemed by the agency to be adequate and equitable. This objective was discussed in the joint statement issued by the Secretary of Labor and the Secretary of HEW on January 2, 1976 (U.S. Department of Labor, 1976).

In conclusion, let me make it clear that the council does not regard the problems of higher education in relation to affirmative action as unique and/or as requiring some sort of favored treatment. In many ways, they are not unique. A government or industrial research department seeking an individual candidate with highly specialized training faces much the same difficulties in measuring the pool of qualified persons as are faced by university departments. If the council's recommendations are concerned exclusively with the problems of higher education, it is because its members were appointed by the Carnegie Foundation to concern themselves with policy issues relating to higher education.

There is one respect, however, in which higher education is unique, and that is that one of its major functions is to provide the educational programs that train the future faculty members of colleges and universities. In view of the extremely deficient supply of qualified members of minority groups in most fields and of women in many fields, this means that, for all practical purposes, colleges and universities cannot achieve the objectives of affirmative action by concentrating on the demand side alone and seeking to eliminate discrimination in hiring and promotion. Unlike other employers, they cannot depend on outside forces to increase the supply of qualified applicants. Thus, in this respect, they clearly have dual obligations that are not shared by most other employers.

REFERENCES

Administrative Conference of the United States. Recommendation 75-2: Affirmative action for equal opportunity in nonconstruction employment. Adopted June 5–6, 1975. Washington, D.C.

Carnegie Council on Policy Studies in Higher Education. *Making affirmative action work in higher education: An analysis of institutional and federal policies with recommendations.* San Francisco: Jossey-Bass, 1975.

National Research Council/National Academy of Sciences. Minority groups: Among doctorate level scientists, engineers, and scholars, 1973. Washington, D.C., 1974a.

National Research Council/National Academy of Sciences. Summary report, 1973: Doctorate recipients from United States Universities. Washington, D.C., 1974b.

U.S. Department of Labor, Office of Information. News Release. Washington, D.C., January 2, 1976.

U.S. President. Manpower report of the President. Transmitted to the Congress, Washington, D.C., April 1974.

Vetter, J. Affirmative action in faculty employment under Executive Order 11246. Draft report prepared for the Committee on Grant and Benefit Programs, Administrative Conference of the United States, May 6, 1974.

23

Rules of the Game: An Essay in Two Parts

JAMES C. GOODWIN

I

The moral arm of the universe is long, but it bends toward justice.

—Martin Luther King, Jr.

Affirmative action is a tool by which institutions of higher education can gauge their excellence. In a large percentage of our universities, there is an enormous gap between the rhetoric of excellence and its reality. Educators have grown uncertain about the social and intellectual purposes of the enterprise; some no longer care enough to give their very best. Our society likes to claim that it is devoted to equality and social change. It has an educational system designed to preserve that contradiction by institutionalizing the rhetoric of change to preserve social stasis.

There is an irresistible temptation to exalt form over substance by concentrating so much on complicated, if not meaningless, systems of data collection, computer printouts, and forms as not only to burden institutions financially but to distract people within them from the really serious, difficult, and ultimately important job of implementation and actual results: "The greatest and most immediate danger of white culture, perhaps least sensed, is its fear of truth, its childish belief in the efficacy of lies as a method of human uplift" (DuBois, 1961, p. 55).

The great white marshmallow of university structure is antithetical to affirmative action. People at all levels of a university system speak

JAMES C. GOODWIN • Assistant to the Vice-President, University of California at Berkeley, Berkeley, California.

readily about the problems of affirmative action, since everyone seems to consider it someone else's problem and disclaims the ability to deal with it on his or her own level. Throughout the nation, affirmative-action offices are being demoted or rendered ineffective as they are moved further away from the decision-making centers within the patriarchal structure and the substructures that make up the university. Through policy and inaction, we see a broken-down charade acted out by the U.S. Department of Health, Education and Welfare (HEW) and the universities, accompanied by the manipulation of data from ineptly tuned computers. The human element has been lost.

In particular, serious, open searches and the objective selection of faculty members have failed to become the rule. Let me share a few examples of resistance.

1. *Smoke Screens.* A young Jewish male scholar with all the requisite qualifications for the position is told that he is the most qualified candidate but that he will not receive an offer. This scholar receives the following letter:

> I am sorry to report that although our department saw you as our top candidate, we will not be able to make you an offer for our new position. Our university is an affirmative-action employer, and the department must attempt to fill the new position with an individual from a recognized oppressed minority group. Although the department initially viewed your ancestry as satisfying the requirements of affirmative action, consultation with our institutional advisers on the affirmative-action program indicated to us that your ancestry does not qualify you as an oppressed minority. I wish you the best of luck in your future, and I am deeply sorry we were not able to extend an offer to you.

In fact, the institution has hired a less-qualified Anglo male candidate from the graduate school attended by the Anglo chairman of the department. The rejected scholar goes to the Anti-Defamation League, which writes to the President of the United States and others, alleging that well-qualified Jews are not being considered since institutions are forced to hire less-qualified women and members of minority groups to conform to affirmative-action requirements.

2. *No Search, Just Seizure.* Dean X has a qualified, but not the best-qualified, candidate whom he would like to hire for a certain high-level administrative position. The dean creates a position at a lower level and then argues that only the person he has previously selected has the

special qualifications to do the job. Even if the position is listed, the personnel office will be unable to refer a successful candidate to Dean X, since the dean has already decided that the person he will hire is the one he has already selected. Once the candidate has been hired, he is swiftly promoted to the higher-level administrative position.

3. *Good-Faith Effort, Game Plan No. 1.* Department chairman Z knowingly offers a position to a minority or woman candidate but under conditions or at a salary level that the particular candidate has indicated would be unacceptable. This looks like a good-faith effort: the candidate was offered the position but turned it down. Chairman Z, having received a refusal from the most qualified candidate, then goes ahead and hires the Anglo male candidate he intended to hire all along—and at a salary higher than that offered to the minority or woman candidate.

4. *Good-Faith Effort, Game Plan No. 2.* A minority or woman candidate is formally listed as the department's second choice after the chairman is certain that the first choice, a white male, will accept the position. This looks good in the recruiting records since it indicates, deceptively, that a minority or woman candidate was seriously considered for the position.

5. *Good-Faith Effort, Game Plan No. 3.* The department informally hires an Anglo male by an oral agreement. Subsequently, as an afterthought and to meet the good-faith requirement, the department recruits (by advertising in the *Chronicle of Higher Education*) and reviews minority and women candidates. Needless to say, these candidates, cynically recruited after the decision to hire the Anglo male, are turned down since the position is, in fact, no longer open.

6. *Good-Faith Effort, Game Plan No. 4, or the Double Play.* A position at university Y is offered to a minority or woman professor at university X. The candidate is informed by the department chairman at university X that the offer is not a "real one based on merit but only an affirmative-action position to meet requirements." He adds, "For you to accept an affirmative-action position would detract from your merit." Thus, the minority or woman candidate is discouraged from accepting an offer, and the offering institution has a good-faith effort for its records. (You may or may not infer prior contact between the two institutions.)

7. *Chickenhearted Scapegoating.* Department chairman Y's former colleague, an Anglo male who is not the most qualified candidate, is let

down easily and dishonestly. The Anglo male is not told that he is not the most qualified candidate; he is simply informed that "we had to hire a minority woman." The conclusion that the Anglo male erroneously draws is that he was the most qualified candidate and that he would have been hired but for the institution's affirmative-action policies. The Anglo candidate complains to his professional association that he has been deprived of a position by the implementation of affirmative action.

8. *Procrastination, or Let-Them-Wait-Another-300-Years.* To defeat the recommendation of the faculty committee to offer the position to a woman candidate, the department chairman makes the offer but then delays the signing of the contract by the date agreed upon. The woman candidate, who has offers with deadlines from other institutions, is thus conveniently forced to remove herself from consideration.

9. *Forked Tongue.* An American Indian with a large educational institution was approached by an acquaintance from another prestigious university who indicated that he was on a search committee for an assistant to the vice-president for administration and wanted to submit the American Indian's résumé. The American Indian replied, "What if I'm not interested?" As they talked, it became clear that the search committee wanted to show that it had approached an appropriate number of minority members and did not care whether the American Indian was, in fact, interested in applying for the prospective position, as long as the résumé was circulated among the members of the search committee to show that a good-faith effort had been attempted. Even though the American Indian did not submit his résumé, he had the uneasy feeling that his name had become part of the applicant pool.

Any minority or women candidates who survive these and other ploys and become members of an academic department are then faced with a whole new series of actions that can result in disablement or removal. But that's another story.

Usually, search committees of academic departments are not instructed in affirmative-action recruitment methods. However, in at least one instance, the search committee recommended a woman to the department chairman as the best-qualified candidate. When the chairman of the search committee reported this result, the chairman was heard to respond, "There is no need to hire her, we already have a woman in the department . . . we have our quota."

The stories in academia about unfortunate and unfair hiring practices are endless. The divide-and-conquer games, playing women against minorities or minorities against women or different ethnic or racial minorities against each other, are common. For example, a department chairman simultaneously promises the same position to women and Chicanos; he promptly steps back and lets both groups fight over it.

Recently, the chairman of a psychology department that is without a woman on the ladder or in a tenured position remarked gratuitously to a woman candidate, "I am not a sexist; ask the girls in the office."

The main features of the "old-boy" method are subjectivity and cronyism. Its network extends beyond the formal walls of the university; for example, members of external science advisory committees are predominantly white, male, and over 35 and are drawn disproportionately from graduate universities. Not surprisingly, their grant awards discriminate against younger researchers and untried approaches as well as against women and minorities. At the same time, these committees perpetuate their biases by long terms of service and the tendency to maintain the entrenched old-boy network by nominating their personal acquaintances and colleagues to succeed them.

Ray Cromley, in the *Berkeley Gazette* of January 22, 1976, reported:

> Recent studies have shed some light on these connections. There seems to be a definite pattern of personnel circulation between certain colleges and universities and certain departments, agencies and bureaus of the government. Graduates and professors regularly move into departments, where friends and associates are located, then back to the university or to special research institutions which have become "holding groups" for scholars out of government, perhaps because of a change of administrations, until the time is ripe to move back into the federal system in higher positions. These men, in turn, bring in their protégés who, more frequently than not, receive rapid promotions.
>
> Sometimes, more commonly today than in the past, the circle is enlarged to include certain major corporations. In the main these companies themselves seem to have been heavily infiltrated by the men from the same academic institutions.
>
> One might suppose offhand that these colleges and individuals were chosen for their excellence. . . . The examples of the work I have seen suggest this is not always so. The favored colleges are not always those with the highest reputations. A goodly number of the

> nation's higher ranking institutions are not represented in appreciable numbers in the upper ranks of the bureaucracy. They are not numbered in the "in" group.
>
> I suggest the growth of these cliques forms an unhappy pattern, dangerously narrowing the base on which government decisions are made.

As the above examples illustrate, institutional racism and sexism are complex in institutions of higher learning. Every day, too many minorities and women are denied the right to seek and qualify for gainful employment or to improve their job status for "reasons" that deny human dignity and are unrelated to any abilities to perform the work in question. There is a great deal of "sociability testing" involved in getting into the professoriate. Criteria not relating to ability creep into the employment processes at crucial decision points, resulting in a self-perpetuating pattern of homogeneity. The pattern will continue unabated until, in Eric Ashby's words, we "reconcile the intellectual detachment essential for good scholarship with social concern essential for the good life" (Ashby, 1974, p. 27). Then we can begin to address the question: What abilities and qualities *are* important and *do* make a difference?

Universities generally admit that they have been less than fair to women and members of minority groups in recruiting, appointing, and promoting them in faculty positions. Before the budget trim and retrenchment, there were many promises to right the wrongs of the past. Then and now, most universities labor under the *a priori* assumption that there are no "qualified" women and minority-group members for academic and upper-level administrative positions. If you don't believe that "qualified" women and minorities exist, then even if they walk into your office, you cannot recognize them, much less engage in bold, innovative approaches to their recruitment and hiring. Also implicit is the assumption that if minorities and women are to be employed, the allegedly high standards now in play will be diminished.

It is not merely racism and sexism that oppose affirmative action but a subtler, more ingrained resistance to recognizing the need for reform. To correct racism and sexism, one must admit that they exist—and in the United States these admissions are not easily made.

Unintentional or respectable racists create code words for a clandestine policy of exclusion and racism. On its face, a word or phrase such as *busing, career education,* and *quotas* is racially neuter and, of course, has many legitimate uses. In effect, these and other code words are used to mask, distort, and corrupt the democratic processes and to titillate the latent racial hostilities of the mass of white Americans. As Jesse Jackson put it, "It's not the bus, it's us." Code words let whites express and vote their cancerous feelings without facing them. They say that affirmative action, instead of giving equal consideration to all Americans, seeks "preferential treatment and quotas" for minorities and women. You hear: "We must hire only the most qualified," or "We can't compromise our standards," or "reverse racism." One can comprehend these phrases only with ears attuned by a lifetime of listening to the language of evasive racism and sexism fed by pervasive fear, misunderstanding, and indifference to the reality of *full and equal employment in jobs that determine the quality of life in an economy planned for, with, and by us.*

Before we get too concerned about discrimination in reverse, we need to deal with the entrenched and pervasive character of racial and sexual discrimination. James Baldwin (1972) observed, "White America remains unable to believe that black America's grievances are real; they are unable to believe this because they cannot face what this fact says about themselves and their country" (p. 165). The effect of this massive and hostile incomprehension ("the arbitrary quality of thoughtlessness") exacerbates the danger for all of us. The more we look upon each other not as human beings but as dehumanized racial or sexual stereotypes, the more firmly closed each group becomes, and the easier it is to hate and destroy.

At the heart of the resentment of many women against their present status is the fact that classifications and distinctions based upon sex are not only discriminatory in themselves but also lend institutional support to entrenched practices that ignore women as persons and treat them, consciously or unconsciously, primarily as sex objects or derivative people, not as full persons. All that has been said about the deprivations and frustrations of women applies with special force to minority women, who have been doubly victimized by the twin immoralities of racial bias and sexual bias.

Walter J. Leonard (1974), special assistant to the president of Harvard, asserts that white male faculties seem "absolutely incapable of developing the internal courage or intellectual capacity to accept other than one of their own kind as equal." There is a point at which elitism becomes indistinguishable from racism and sexism.

Racism and sexism refer not only to the bad acts of bad people but also to well-established and entrenched patterns of institutional behavior that are neutral on their face but, intended or not, result in the reinforcement and maintenance of present inequalities stemming from past discrimination. The objective is to ensure not only that all Americans play by the same rules but that all Americans play *against the same odds*.

Perhaps the cruelest aspect of the current cry of "reverse discrimination" is that it ultimately deprives women and members of minority groups of the satisfaction of knowing that they have made it on their own merit. Racism is a system of deadly oppression, both spiritual and physical; that system is not reversed by isolated instances of discrimination against Anglos or by the favoring of one minority over another. Regardless of whether reverse discrimination is wise or unwise, it is impossible for a majority, any majority, to discriminate against itself.

If we are to develop new ideals and grant them some autonomy from our narrow empirical reality, we will have to draw on more diverse cultural, social, economic, and political models than those now provided. Universities need people in key positions with creative problem-solving skills to develop mechanisms for balancing sexual and racial diversity with quality, in keeping with the values of cultural pluralism and bilingualism.

In a recent issue of the *Chronicle of Higher Education*, there was a comment about Harvard's economics department, concluding with the following: "In the final analysis, too abstract a theory, too narrow a focus on one methodology, too uncritical an acceptance of economic laws or too specialized a structure of the discipline may all have contributed to the failure of economics to predict the present strange combination of inflation and recession" (Winkler, 1975, p. 1). What is clearly indicated here are the kinds of myopic questions asked and problems raised that creep incestuously into an inbred system that has lost its way. Thus, homogeneity, overspecialization, and conservatism converge as heavy obstacles to weigh down the pursuit of excellence.

I criticize institutions of higher education because I want to raise their level of moral dignity. A university that does not make a major, persistent effort to reflect our pluralistic society in all its departments, among its faculty, and in its student body fails to recognize that the university should serve as a model. If the university will not voluntarily assume the role of a pluralistic model, we must compel it and by doing so provide a space for a guerrilla theater, right in the vitals of the military–industrial–academic complex. The educational institution is society's Achilles heel, and while the war may not be won there, it may be lost—pluralism is the constitutional answer to insensible power.

Immanual Kant (1960, p. 66) held that all of our activity as rational beings is focused on three questions: What can I know? What ought I to do? What may I hope? Universities have traditionally placed emphasis on the first of these questions: What can I know? The university must rise to the challenge of its responsibility on the uses of knowledge and the consequences of knowing. If its primary process is inquiry and its primary product is knowledge, its primary concern must be the human intellect and its ultimate concern the human being and the future of humanity. The university must embrace not only the first of Kant's questions: What can I know? but the second and moral question as well: What ought I to do? Morality is the search for justice. A famous citizen of Athens once was asked when he thought justice could be established. He replied, "When those who are not injured feel as indignant as those who are."

II

Where does the agency for democratic change reside in the university?

Too often, unfortunately, the price of reasonable progress in this country has been court-enforced accountability (whenever it has occurred)—the dialectic of progress and coercion. We seem to make progress in institutional life only through the conflicts and coerced changes brought upon the university by minorities and women, by students, by the community, and by state and federal governments.

Of course, graduate and undergraduate admission policies for minorities and women are linked to employment practices; educational opportunities and employment opportunities are indivisible. If an employer or admissions officer uses reasonable flexibility to help remedy the racial or sexual imbalance resulting from past exclusion, she or he is not necessarily guilty of discrimination in reverse. On the contrary, it can be strongly argued that she or he is fulfilling an ethical obligation to reexamine and improve selection criteria that are racist or sexist in effect, if not in intent.

Some years ago, the most prestigious eastern universities recognized that as a result of regional differentials in secondary education, few Southerners, Midwesterners, or other "provincials" could be expected to filter through the admissions screen; consequently, these universities waived college-board examinations for the non-Easterners and substituted the criterion of the applicant's high-school performance—a straightforward example of "preferential treatment" that no one was heard to complain about. Instead, at that time, although the beneficiaries of preferential treatment were overwhelmingly white and male, this effort to achieve a pluralistic result was considered laudably democratic. The important point is that the universities involved presumably recognized the value of pluralistic student bodies in the growth of students and in sound institutional development. And since education at a prestigious university has a decidedly favorable influence on later career opportunities and advancement, these institutions had a "democratizing" effect on business and professional leadership for the benefit of the society as a whole.

Meanwhile, with not enough concerned faculty and students tuned in, the prospects for achieving affirmative action are dim. Our task is heavy: to educate, organize, mobilize, and coordinate an effort to establish firmly the principle that affirmative-action programs benefit the quality, the objectivity, and the fairness of a university's academic and staff personnel system by increasing the pool of available qualified candidates and applicants under higher standards of quality and productivity. Such a pool would also reflect more accurately and appropriately the diversity of our society throughout the university work force.

The purpose of affirmative action is to raise standards, not to lower or maintain them. Standards are raised not only by an honest, broad, innovative search that includes all the competent minority and women candidates available but by the establishment of cultural pluralism in our faculties. The two purposes—increased productivity and greater equality—are inextricably joined.

A government committed to not subsidizing discrimination among its contractors cannot rightfully exempt contractors simply because their business happens to be education. Experience with nondiscrimination and affirmative-action regulations, state and federal, has overwhelmingly shown that little or nothing happens so long as institutions are not held accountable for measurable results.

Universities, because they create and cultivate their own applicant pools, can develop their own goals and timetables. We have in our own student bodies the very stuff it takes to make the kind of culturally diverse faculty and staff we so badly need. Even Caspar Weinberger (1975) agreed, "This is particularly true because of the fact that colleges and universities for the most part control the access of persons to the academic employment pools from which they recruit."

Diversity enhances excellence. Students, the constituency that distinguishes a university from other institutions, are the vital but frequently overlooked beneficiaries of a university's affirmative-action program. Broad representation of qualified women and minorities within faculty, staff, and student body improves the quality of university education. These educational priorities dictate that universities are obligated to their students to provide effective role models, support, and counseling. In her study of the effects of women faculty and men students on the subsequent career success of 1,100 women college graduates, Dr. Elizabeth Tidball (1973), professor of physiology at George Washington University Medical Center in Washington, found that:

> as the women faculty/women student ratio increases, so too does the output of career-successful women graduates. Conversely, the higher the percentage of men students enrolled, the smaller the output of women achievers. The proportion of women faculty in all undergraduate educational institutions has been declining for the past forty years, while the enrollment of women students has accelerated markedly in the last decade, especially in the coeducational colleges.

Stanford Cazier (1972), President, California State University at Chico, stated:

> Augmenting this representation can bring to our campuses a variegated set of experiences, perspectives, reflexes and sensitivities that are vital to an understanding of and ability to cope with a rapidly changing world. Not only do ethnic minorities need the availability of faculty members with whom they can relate and to serve as models for them, but the middle-class white student deserves exposure to cultural nuances which ought to inform the cognitive and affective baggage of any liberally educated person. (pp. 5–6)

The model of a graduate 40 years ago, wrote Eric Ashby (1974), was the "person ready to take responsibility for preserving a set of values which he felt no mind to question. . . . That sort of person cannot cope with the modern world. The contemporary model is a person educated for insecurity, who can innovate, improvise, solve problems with no precedent" (p. 7). Education is characterized by a belief in the reality of higher human needs, motives, and capabilities than have previously been acknowledged. Diversity and quality are compatible and essential.

The federal government has not mandated the maximum feasible participation of minorities and women in affirmative-action programming; both HEW and the universities have insufficient goals and timetables for student, staff, and faculty. Mature struggles are not built on narrow interests. Neither are they built solely upon psychobiological needs; indeed, they are built on needs that supersede caste and parochial group interests. They arise out of oppression that cuts across race, ethnic, class, and sex lines and creates out of these disparate and broad groupings a new social force set afoot by a new man and new woman—together.

In short, affirmative action speaks to governance—to a broader, shared authority in the decision-making and policy-formulation processes. (If opportunities are offered without a sharing of power, then we have paternalism.) Before the routinely called-for "reordering of priorities," there must be a reorganization and reordering of power relationships in the university. To expect substantial increments in the employment of women and minorities in a less than steady-state job market without a readiness to change the structural alignment of decision-making and policy-formulation processes is to invite cynicism and disillusionment.

The question of affirmative action is one around which the active citizenship of students, minorities, women, and junior faculty members in the university community can be redefined. The "old boys" simply cannot cope with the new breed, and this includes not only blacks and women and other minorities but many Anglo men who do not fit the old-boy stereotypes. Of course, those of us who want to change the university must first understand it. In the serious process of comprehending the university, we will also see what the university is not, and what potentially it might be. What "is" and what "ought to be" are inseparable; but the "ought" can be realized only in action.

It is no startling psychological insight that most of us are delighted to hear and prone to accept characterizations of ethnic or other groups that suggest that our own group is superior to theirs. For instance, groups that have struggled long to gain advantages do not readily yield the fruits from their own struggles so that others may share them. Thus, the principle that we can keep our abundance only by sharing it is neither appreciated nor understood; and yet no advantage is safe, no gain is secure unless offered to all alike.

The task ahead is both large and unpopular; nonetheless, it deserves the greatest energies if institutions of higher learning are to have a meaningful role in the evolution of a more humane society. I still believe that racism and sexism are the major unfinished business of America, that there can be and sometimes is a beloved community of resistance, that there is an American revolutionary morality and tradition (from founders like Tom Paine, Abigail and Sam Adams, to abolitionists like Bill Garrison, Fred Douglass, the Grimke sisters, and Sojourner Truth), and that it is from that tradition that we take our sustenance. For if we do not save the best in our heritage and make it live in the present, then we will have no future. Where does the agency for democratic change reside in the university? Where else but in our hearts.

REFERENCES

Ashby, E. *Adapting universities to a technological society.* San Francisco: Jossey-Bass, 1974.

Baldwin, J. *No name in the street.* New York: Dial Press, 1972.

Cazier, S. Accolade for affirmative action. Address given to faculty and staff on March 14, 1972.

DuBois, W. E. *The souls of black folk.* New York: A Crest Reprint, Fawcett, 1961.

Kant, I. *Education.* Ann Arbor: University of Michigan Press, 1960.

Leonard, W. Public forum on affirmative action: A philosophy and program for justice. Keynote address at University of California at Berkeley on June 1, 1974.

Tidball, M. E. Perspective on academic women and affirmative action. *Educational Record,* 1973, *54*(2), 130–135.

Weinberger, C. From an address delivered by the former Secretary of HEW to the Commonwealth Club of California in San Francisco, July 21, 1975.

Winkler, K. J. Economists are asking: What went wrong. *Chronicle of Higher Education,* 1975, *10*(13), 1.

VI. Special Training Programs for Minority Students in Science: College Level

24

Prehealth Summer Programs

RICHARD P. McGINNIS

Tougaloo College has, for the past four years, been engaged in a special project to increase the number of minority students from Mississippi who enter the health professions and return to the state to practice. Tougaloo is a small, predominantly black liberal-arts college with an average enrollment of 720 students. The college draws 90% of its students from Mississippi.

The program that we have developed to direct students into the health professions has several components. These include visits to high schools by Tougaloo science students to encourage other minority students to pursue careers in science and the health professions; a summer program emphasizing the skills necessary for college science courses and providing career counseling for 45–65 entering freshmen; a health-careers club; a preceptorship program with local physicians, hospitals, and medical schools; and counseling and assistance in entering health-professional schools. Tougaloo students also participate in special summer programs sponsored by Harvard, Baylor, Princeton, Meharry, Yale, and Fisk universities.

The six-week summer program for about 50 prefreshmen students addresses the academic difficulties that many of our students face. These generally are a lack of math skills; inexperience in quantitative problem-solving; a lack of exposure to science curricula in high school; problems with communication skills; and inadequate knowledge of career opportunities in science and the health fields. Consequently, our program features noncredit courses in math, chemistry, problem solving, biology, and English (titled "Scientific Thinking and Scientific Communications").

RICHARD P. McGINNIS • Associate Professor of Chemistry, Tougaloo College, Tougaloo, Mississippi.

The focus of the course work is to develop analytical tools that will be applicable to all science disciplines. Thus, the chemistry course, while furnishing the rudiments of chemical vocabulary, concentrates on the translation of word problems into equation form and on computational accuracy. The biology course stresses the use of library facilities and acquaints students with laboratory apparatus. The math course, offered for the first time in 1975, reviews algebra and geometry. By special request, one group of students studied calculus.

The English course has gone through a good deal of evolution. Initially, it was an intensive writing course. However, we have also experimented with elements of logic, vocabulary building, and science reading. This past summer, we asked students to write concise summaries of scientific articles at the *Scientific American* level, and the assignment turned out to be quite challenging to them.

In addition to academic skills, the program also provides the students an orientation to health careers (meetings with health professionals and curriculum planning) and an introduction to the social aspects of health policy. Most importantly, through participation in the prefreshmen program, most of the students gain confidence in their ability to pursue a health curriculum and begin to recognize the effort required for success in the science professions.

In the summer preceptorship program we try to emphasize the health-care delivery problem in Mississippi. In this effort, we are aided by many physicians, dentists, and allied health professionals near the students' homes. Students receive firsthand exposure to activities in professions about which they have a very limited knowledge (example: dentists do more than pull teeth!). Black health professionals provided role models for the students. Students work for 8–10 weeks at hospitals, medical schools, comprehensive-health centers, and private clinics. Stipends are paid through the program. In general, the students have enjoyed the preceptorship program and feel that it has motivated them to do better academic work. The difficulty, of course, is finding suitable placements for students in many areas where health care is very inadequate.

Both programs are having a sizable impact on the college. We now have more students interested in science, and they are better motivated and better prepared than ever before.* Enrollments in science curricula

*The science enrollment for the past four years is as follows: 1972–1973, 150; 1973–1974, 168; 1974–1975, 254; 1975–1976, 292.

have more than doubled in three years. More than 50% of this fall's entering freshman class of 220 students planned to enter a health career, up from 10% a few years ago. The grades of the students who participate in the two above-mentioned programs are about 0.5 grade points above those of other science students. Scores on standardized examinations (General Chemistry—American Chemical Society) have risen also, from a mean of 12th percentile to 28th percentile. This improved academic performance should reflect itself in increased admissions to medical, dental, and graduate schools in 1977, when the first participants in the prefreshmen program will apply.

However, many problems remain. Much more work is needed in the area of preparation of students for standardized testing. New methods of teaching the skills of test taking should be developed for these students. All of these activities should carry over into better science problem-solving ability.

25

Basic Science Enrichment Courses for Minority Medical Students

ALONZO C. ATENCIO

As one examines the figures reflecting minority representation in medicine, dentistry, law, science, engineering, and the related professions, one is struck by the obvious exclusion of minorities from these areas. In medicine, for example, if we were to achieve the national average of 600/1 patient-to-physician ratio, we should have at least 40,000 black physicians instead of some 6,000 now listed in the National Medical Association (NMA) rolls. Figures for the number of Chicano and American Indian physicians are so small that they are statistically insignificant. For the present population of Chicanos, we should have at least 20,000 physicians to reach the 600/1 ratio.

Other figures compiled by the National Science Foundation (1975) also reveal gross underrepresentation of minorities in the fields of science and engineering. These figures show that there are only 4.3 black scientists per 10,000 black population compared to 26.5 per 10,000 white population. In engineering, there are only 2.9 blacks compared to 45.7 white engineers per 10,000 population. Again, the figure for Chicanos is so small that it is statistically insignificant. The Society for Advancement of Chicanos and Native Americans in Science, for instance, has identified less than 100 Chicanos and 16 American Indians in science at the Ph.D. level. Yet, there are at least 1,283,000 scientists and engineers (NSF, 1975) in this country. These figures clearly verify our exclusion from the major professions. Today, I shall attempt to analyze some of the reasons leading to the exclusion of minorities. The process of exclusion can be visualized more clearly in a flow diagram showing the flow of minorities through the educational pipeline.

ALONZO C. ATENCIO • Assistant Dean, University of New Mexico School of Medicine, Albuquerque, New Mexico.

This flow of students, which represents 100% at the beginning because of compulsory education, begins at the elementary-school level. They continue through the intermediate- and secondary-school system on toward college, master's degree programs, and professional school. In Figure 1, the flow through the system is indicated by arrows, in this instance, toward medical school. The arrows leading away from the boxes represent leaks from the system into boxes labeled "losses." The box for minority losses is larger, to show that the losses are much greater for the minority.

The magnitude of these losses is shown graphically in Figure 2, which shows the trends in the Southwest indicating the mobility of the minority and the majority student. Note that the majority population in the Southwest is initially 68%, as opposed to 18.5% for the Chicano,

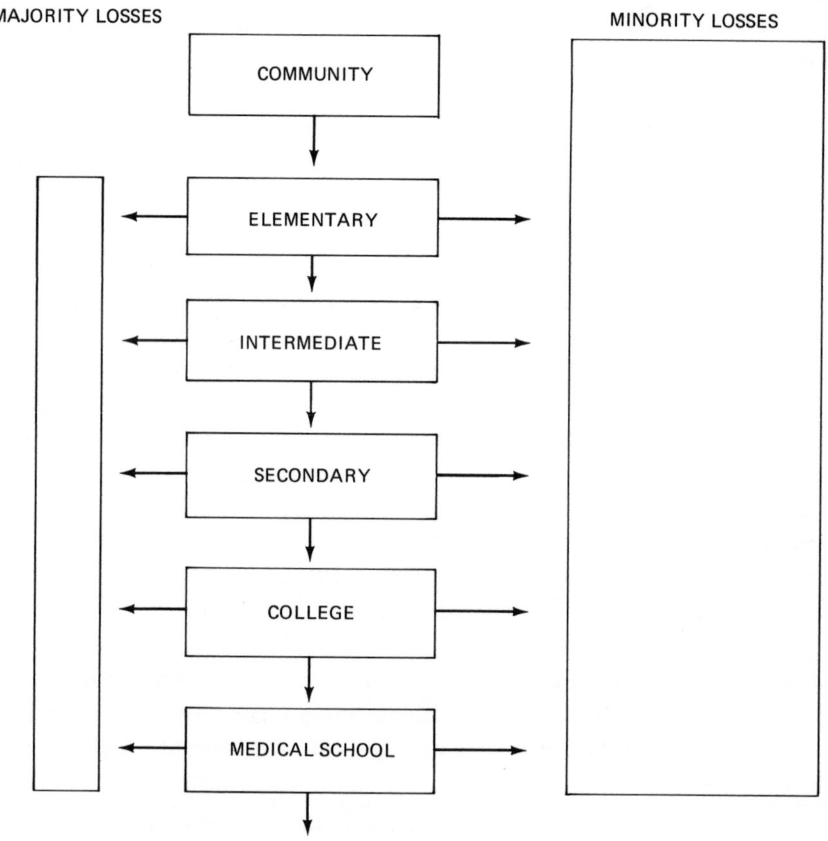

Figure 1. Flowchart of the educational pipeline.

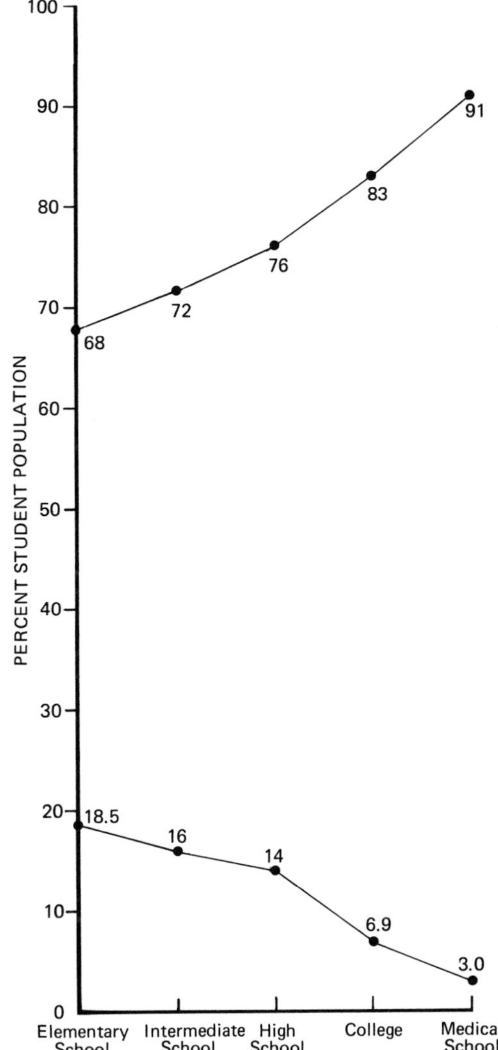

Figure 2. Educational trend in the Southwest.

and is well into the millions. As the student from the elementary school moves on toward intermediate school, the percentage rises from 68% to 72% for the majority student, while the Chicano student population drops from 18.5% to 16%. In high school, the Chicano student population has dropped to 14%, while the majority has increased toward 76%. As students approach college, the flow of Chicano students has plummeted to 6.9%, but the majority has now risen to 83%. At the medical-

Table 1. Population-Distribution Differential in the Southwest

Institution	Anglo	Chicano
Elementary school	0	0
Intermediate school	+4.0	−2.5
Secondary school	+8.0	−4.5
College	+15.0	−11.6
Medical school	+23.0	−15.6

or professional-school level, the disparity is obviously much greater. The majority is in excess of 90%, while the Chicano student population has dwindled to a mere 3%. The figures are for the Southwest only and do not reflect the national distribution. For example, there are only 640 Mexican-American students currently (1975) enrolled in medical school (AAMC, 1976). Nationally, this number is closer to 1%, rather than 3% as in the Southwest.

The divergent trends are more clearly visualized in Table 1, which presents the differential increases and decreases in the data. If the net change at elementary school is taken as zero, the Chicano has a net decrease of 15.6% and the majority (Anglo) has a net gain of 23%. The divergence is now 38.6% between the two groups.

Figure 3 shows a representative flow of students through the educational system for the state of New Mexico, where the population is 49% Anglo, 40% Chicano, and 7.5% American Indian. The trend is similar to that presented for the Southwest; the Anglo representation rose from a 49% total population to over 90% in medical school by 1969. The Chicano population, however, dropped from 40% to 4% during the same period. The trend is very similar for the American Indian. The dashed line represents the change in the trend because of innovative minority programs at the University of New Mexico School of Medicine. (A more thorough description is given elsewhere in this paper.) The differential increases and decreases in New Mexico are shown in Table 2.

Table 2. Population-Distribution Differential in New Mexico

Institution	Anglo	Chicano	American Indian	Black
High school	+4.0	−3.2	+0.2	−0.5
College	+26.0	−20.9	−5.6	−0.3
Medical school	+35.0	−36.0	−6.2	−0.7

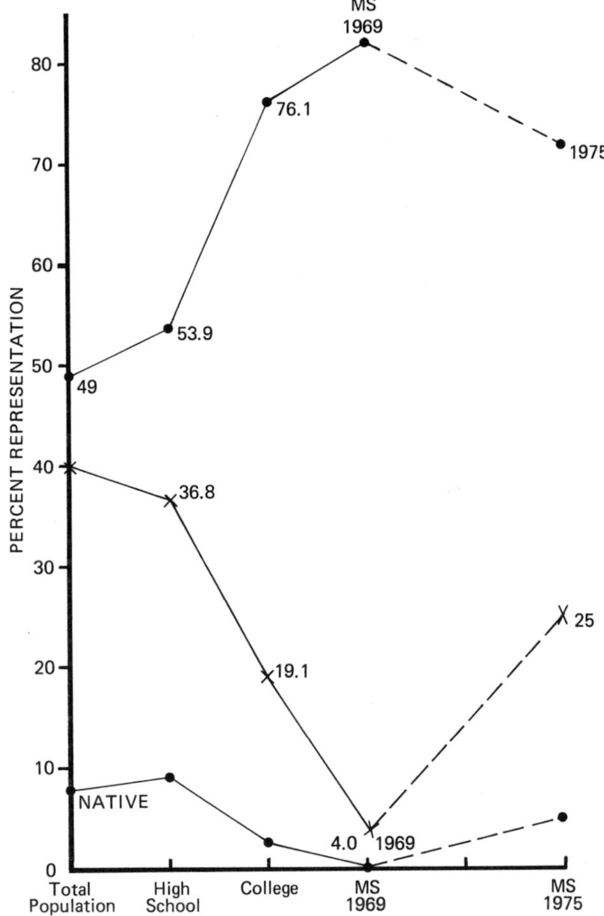

Figure 3. Educational trend in New Mexico.

The presentation so far has been concentrated on the description of the problem. The losses and mobilities can be calculated and quantitatively described by linear differential equations. The arrows here would represent rate constants. The model, therefore, can be described and fed into a computer. The conditions leading to the losses, however, are much more complex and need to be enumerated and understood if the necessary action is to be undertaken to produce changes.

Let us examine the characteristics of the system that produce leaks of minorities from the educational process. What causes these leaks? Why are Chicanos more permeable to the educational pipeline?

The answers to these questions are complex and many causes have been cited. The citing of causes such as the absence of role models, poor counseling, the irrelevance of the educational system, racism, and so on down the line is too simplistic. As one begins to analyze the situation in general terms, the process becomes repetitious. For example, cultural conflicts and poor counseling are found at all levels of transition when one applies them to monolingual students from a different culture. The varying degrees of intensity of these factors at each stage are not considered. Nor is the intensity of the impact of these factors at individual steps of development and transition accuratedly described.

Let's consider the transition of children from the general population at the age of 6. School attendance is compulsory; the students come primarily from monolingual parents who are not widely read in contemporary literature; the children are highly impressionable and eagerly anticipate active participation in school and its learning process. It is at this stage that the lack of role models and the parents' lack of a positive view of education are often cited as the primary contributors to the students' performance in school. Somewhat more realistically, however, the proponents of bilingual education cite the lack of bilingual programs and teachers as major contributors. Yet, many monolingual foreign students adapt and do well in this educational system, so the reasons for the disparity in performance must lie elsewhere. The U.S. Commission on Civil Rights (1972) has cited some concrete examples that show that the general attitude of teachers toward Chicanos in school is perhaps the most important factor at this stage of development. For example:

> One Chicano sat toward the back in a corner and volunteered several answers. At one point the teacher did not even acknowledge much less reinforce his answer. At another time, he volunteered an answer which was perfectly suitable, yet the teacher states, "Well yes, uh huh, but can anyone else put it in different terms?" . . . the teacher then called on an Anglo student who gave the basic response with very little paraphrasing. The teacher then beamed and exclaimed, "Yes, that's it exactly."

They recorded other observations of classroom behavior and I will cite two more:

> There were several Chicanos who kept raising their hands eagerly at every question. Mrs. G. would repeatedly look right over

> their hands, and call on some of the same Anglo students over and over. In some cases she would call on the Chicanos only because the Anglos stopped raising their hands. After a while the Chicanos stopped raising their hands.
>
> Mrs. M. was leading a class discussion on unions, but all the interaction was between the teacher and three Anglos sitting in the front of the class. They were very eager, but the rest of the class was bored. Mrs. M. finally said, "The same hands, I always get the same hands."

In another report, entitled "Teachers and Students," the U.S. Commission on Civil Rights (1973) concluded that the schools in the Southwest are failing to involve the Chicanos as active participants in the classroom. On most measures of verbal interaction between teacher and student, they found gross disparities that favored the Anglo student. Briefly, they found that (1) the teachers praised or encouraged the Anglo children 36% more than they did the Mexican-American children; (2) they built upon contributions of Anglo pupils 40% more frequently than on those of the Chicano; (3) they directed questions to Anglo students 20% more often; and (4) overall, the Chicano child received less attention from the teacher. In short, they found that the Chicano child was ignored more than the Anglo student.

These examples and others have been documented, but I can reinforce them from my own experience in coping with an educational system that has and is failing to educate our young people. The treatment described above causes serious damage to the young Chicano's ego. Obviously, what I am saying here applies to other minorities as well. In fact, the situation is much more serious with the American Indian. In addition, the humiliation and the segregation of the black community has been well documented. Dr. Harvey Webb (1973) has written a well-documented report on the barriers blacks encounter in gaining access to health professions.

In such classroom settings with limited interaction between teacher and student, how can a climate of learning that directly affects their educational opportunities be created for the Mexican-American child? Could it be that this is one of the reasons less than 20% of the Chicano high-school students aspire to a college education?

The transition from high school to college is fraught with other barriers arising from American College Testing (ACT) tests and admission policies. The performance on the ACT presumes to measure the fund of knowledge a student has acquired in high school. The tacit assumption here is that all the students are comfortable with standardized tests and that the only variable is the content of the examination. Of course, there are other variables that contribute to the students' performances—assuming that the quality of the high-school education is good. This assumption, however, is not always true, for the quality of barrio and ghetto schools is well known to us.

Another interesting concept was brought out recently by Fernandez (Espinosa, Fernandez, and Dornbush, 1975), that is, the concept of miseducation, which differs from the overt discriminatory attitude of the teachers toward Chicano students. These investigators examined the premise that minority parents hold education in low esteem and therefore fail to encourage their children to pursue an education. These authors discovered that this assumption is a myth. Interestingly, they showed that the Chicano student's academic performance did not fit the image the students or the parents held toward education. The majority of the Chicano parents and students believed that there was a close link between school and future occupation, yet Chicano students emerged from high school with relatively low verbal and mathematical skills. The consensus of the authors was that the teachers failed to set challenging standards for these students; consequently, the letter-grade average did not reflect what the student had gleaned from the course. Therefore, the students with above-average grades held a false view of their own skills. This faulty assessment helped perpetuate the image that the Chicano student was inferior. In sharp contrast to the problem concerning teachers with an overt discriminatory attitude, the authors found that the same negative result was obtained with warmth and friendliness; in this case, the student was misled into believing that he was doing well in substandard tests. The students' performance on the ACT, however, revealed a more realistic picture of the students' acquired knowledge. Low ACT scores, in addition to having a devastating effect on the students' self-image, often lead to the denial of admission to college and thereby to an opportunity for further education.

In my experience, minority students rise to the occasion when serious but fair challenges are set forth. In this day of vigorous competition, miseducation is a cruel joke and can lead to irreparable damage that may affect a student's entire life.

There is not sufficient time to go into a more detailed examination of the problem; suffice it to say that the presentation thus far gives one a pictorial summary of the problem. These two examples alone, I believe, are applicable to all levels of education encountered by minority students. Obviously, any attempts to correct the imbalance in minority representation should address the lower levels of the educational system as well as the more sophisticated ones at the professional schools. The patterns as well as the methods and attitudes of teachers are essentially the same. Therefore, any corrective measure suggested should address all components of the educational system.

Details of programmatic design are beyond the scope of this paper. The summer program at the University of New Mexico School of Medicine, however, brings out some of the elements necessary to produce immediate change. However, the long-range solution still remains in the realm of the lower educational levels. The Basic Science Enrichment Program (BSEP) at the UNM School of Medicine is starting its sixth successful year. Its primary goal has been to increase minority representation by encouraging medical schools to accept more "high-risk" students. A high-risk student is defined as a student who has demonstrated academic potential but has low paper credentials, that is, a low grade-point average and low Medical College Admission Test scores. These students are usually denied admission. One of the qualifications of the participants in the BSEP is confirmation that the medical school has offered the student a place in the incoming medical class.

The essential components of the program center on building confidence by having the students participate in a summer course consisting of human anatomy, medical biochemistry, physiology, and medical sociology. The courses are taught by medical-school faculty as rigorously as those taught to regular medical students. The mechanics of teaching include lectures, small discussion groups, peer tutorials, laboratories, and periodic examinations. On a ratio of 4 to 1, the students are assigned a tutor who is a minority medical student. The tutors, besides providing academic counseling and tutorial help, also pass on academic and social-survival skills to the students.

Minority students applying to medical school bring with them many of the hang-ups imparted by the educational system. As a rule, they have low confidence in their ability resulting from some of the attitudes of teachers and from miseducation. In the BSEP, the challenge is high, and as they progress, they learn to cope with the course material and also to thrive in this challenging situation.

Many of the students—for instance, those who remain at the UNM after their matriculation—feed into the secondary-school system as visible role models for high-school students. In high schools and colleges, they give lectures, conduct seminars, recruit, and have contact with many premedical students, encouraging them to develop their skills and to enroll in professional school. In this manner, they counteract much of the negative influence of high-school counselors and teachers, while providing visible evidence that professional school can become a reality for others as well.

There are other summer programs aimed at short-term reduction in losses of Chicano students from the educational pipeline. For example, the University of New Mexico Cultural Awareness Program is designed for counselors, teachers, and high-school administrators. This particular program concentrates on an awareness of the major cultural differences between Chicanos and Anglos and how misunderstanding of these cultural differences often leads to stereotyped conclusions.

Without the presence of dedicated faculty members and administrators, many programs will fail, for they are perhaps the most important ingredient in successful summer programs. Without these dedicated individuals, many programs become mere reproductions of the existing educational system.

REFERENCES

Association of American Medical Colleges. Office of Minority Affairs. Division of Student Programs. Minority Student Opportunities in U.S. Medical Schools, 1975–76

Espinosa, R., Fernandez, C., and Dornbush, S. Factors affecting Chicano effort and achievement in high school. ATIBOS #1. *Journal of Chicano Research,* 1975.

National Science Foundation. Science resource studies highlights: Racial minorities in the scientist and engineer population. September 19, 1975.

U.S. Commission on Civil Rights. The excluded student, Report III: Educational practices affecting Mexican-Americans in the Southwest, 1972.

U.S. Commission on Civil Rights. Teachers and students, Report V: Mexican-American education study, differences in teacher interaction with Mexican-American and Anglo students, March 1973.

Webb, H. Barriers to blacks in dentistry: Barriers to increased representation of non-whites and women in the health professions. U.S. HEW/NIH Contract #NO1MB34052. Baltimore: Pagan and Morgan Associates, 1973.

26

Undergraduate Research Training

FRANKLIN D. HAMILTON

In 1971, the staff of the University of Tennessee–Oak Ridge Graduate School of Biomedical Sciences, recognized that minorities as a group were underrepresented in the biomedical sciences. A program was initiated to increase, in a significant fashion, the number of black Ph.D.'s in the biomedical area. However, before I describe the details of the program we have designed in Oak Ridge, let me first give you some background information about the graduate school, which is a unique institution for graduate education in the biomedical sciences.

The Graduate School of Biomedical Sciences was established as a joint effort between the Biology Division of the Oak Ridge National Laboratory and the University of Tennessee at Knoxville to allow the exceptional talent and research facilities of the laboratory to be more fully utilized in the support of scientific education. The Oak Ridge National Laboratory is one of the many research institutions of the Energy Research and Development Administration. Since the establishment of the graduate school and after the arrival of its first class of seven students in 1967, it has grown and developed steadily and has established a reputation as a center of quality graduate education in the Southeast. The Graduate School of Biomedical Sciences currently consists of 5 full-time faculty members and approximately 80 shared faculty members who are on the staff of the Oak Ridge National Laboratory. Currently, the school has an enrollment of 44 full-time predoctoral and 32 postdoctoral trainees. Of the 38 students who have received Ph.D. degrees from the training program, a sizable number have advanced beyond their postdoctoral studies to obtain permanent appointments in research institutes or universities across the nation.

FRANKLIN D. HAMILTON • Associate Professor, The University of Tennessee–Oak Ridge Graduate School of Biomedical Sciences, Oak Ridge National Laboratory, Oak Ridge, Tennessee.

Very early in the school's history, there was a recognition of the lack of an adequate representation of minority groups in the biomedical sciences. The director of the graduate school, at that time Dr. R. C. Fuller, felt that the school and the biology division, located in southeastern Tennessee in close proximity to a large number of the black undergraduate colleges, would be a logical place to undertake the training of black students in the biomedical sciences. Aided by the efforts of many individuals at both the University of Tennessee and the Oak Ridge National Laboratory, including Dr. Alexander Hollaender, former director of the biology division, the staff of the graduate school successfully initiated a program called "Aid to Black College Students and Faculty Interested in Careers in the Biomedical Sciences." The program, supported largely by grants from the National Institute of General Medical Sciences and the Carnegie Corporation of New York, supports a select number of black undergraduate students and faculty members from black colleges for research training in the Biology Division of the Oak Ridge National Laboratory. In addition, it provides for the training of black students at both the predoctoral and the postdoctoral level in the graduate school's training program.

Since the inception of the minority training program at Oak Ridge, there has been a recognition at the national level of the need for greater training efforts in the biomedical sciences for persons of minority backgrounds. This recognition was brought about in part by the Carnegie Commission report on the plight of black colleges in their struggle for survival (Carnegie Commission on Higher Education, 1971). Within the past few years, programs for the training of minorities in the biomedical sciences such as Minority Access to Research Careers (MARC) and the Minority Biomedical Support program (MBS) have been developed within the National Institutes of Health.

These programs have been very beneficial in stimulating an awareness among black college students of career opportunities in the biomedical professions. However, of the more than 60 minority institutions that have received funds from the programs, only 4 are approved to offer Ph.D. degrees in the physical or natural sciences. Therefore, if there is to be a sizable increase in the output of black Ph.D.'s in these areas, majority institutions must become involved in working toward this end in a meaningful way. Unfortunately, minority enrollment in the graduate schools of majority institutions is still considerably lower than the minority representation in the general population (approximately 12% as determined by the 1975 census). The current enrollment of black students in all graduate programs is 4.4%. Of that number, however, the black enrollment is only 1.2% in engineering, physical, and natural sciences (National Academy of Sciences, 1976). The number of blacks who hold Ph.D. degrees in the natural sciences has increased from 600 in 1969 (Jay, 1971) to approximately 1,155 in 1975 (James Jay, in this volume). Although this number represents an increase of 50% in a six-year period, the total numbers are still less than 1% of all Ph.D.'s in the natural sciences. The need for a sizable increase in the number of black holders of Ph.D.'s in science and engineering is eloquently addressed in a recent report by Dr. Adolph Y. Wilburn (1974). He suggested that the nation's graduate schools should strive to produce 900 black science Ph.D. candidates per year by 1980 if we are to reach parity in the future. In an effort to contribute to this objective, the program at Oak Ridge is designed to take advantage of the outstanding training opportunities at the Universities of Tennessee–Oak Ridge Graduate School of Biomedical Sciences to prepare black undergraduate college students for careers in the biomedical sciences. An integral part of this program provides for the training of these students at the predoctoral and postdoctoral levels. It is the major long-range goal of this program to develop a sufficient pool of professional talent to meet both regional and national demands.

UNDERGRADUATE TRAINING

In the program at Oak Ridge, undergraduate students in predominantly black institutions are given an opportunity to obtain research training for 10 weeks during the summer session in the Biology Division of Oak Ridge National Laboratory. Students majoring in chemistry, biology, and physics are invited to apply. Students are recruited for the summer program through visits to the predominantly black institutions by the graduate-school staff. All individuals who express interest in research are encouraged to apply. The final selection of students is based on academic records, letters of recommendation from their faculty advisers, letters of interest, and personal interviews, where possible.

Students are generally nominated to participate in the program during their junior year of college, but outstanding sophomore candidates may be selected to participate. The selection of a student in the sophomore or junior year presents an opportunity to spend more than one summer of research training at Oak Ridge National Laboratory. Selected students are invited to the biology division to work on research projects under the supervision of biology-division staff members. While at Oak Ridge, the students receive laboratory-research experience, course work, and information on graduate education.

Students work 30–40 hours a week in the research laboratory under the supervision of second- or third-year biomedical graduate students or with staff members of the biology division in such areas as biochemistry, cell biology, mammalian and microbial genetics, biophysics, and physiology. Upon completion of the 10-week training period, the students participate in a formal seminar program in which each student gives a 10-minute presentation covering the scope and results of the research project that he or she has undertaken for the summer. In addition to the research training, students spend 5–10 hours a week in the following courses.

Cell and Molecular Biology. Available to all students. A course designed to cover cell structure and functions at both the molecular and biochemical levels. Topics covered are membrane structure, mitochondria, chloroplast and nuclear functions, and biosynthesis of macromolecular cellular components such as DNA, RNA, and proteins. Texts are *Molecular Biology of the Gene,* J. Watson (1976), and *Biology of the Cell,* Stephen Wolfe (1972).

Introductory Laboratory Techniques in Biochemistry. Required of all first-year summer students. A laboratory course designed to introduce students to techniques required in a research laboratory. Students perform a series of experiments centered on comparative enzymephysiology and tryptophan metabolism in mammals and fungi. Students are acquainted with fundamental biochemical laboratory techniques, shown the necessity and importance of accurate record-keeping, and given exposure to the concept of the scientific method. Texts are supplied by the instructor.

Introduction to Electron Microscopy. Available to all students. A course designed to introduce the students to the electron microscope as a tool for scientific investigation. Three basic techniques of sample preparation are taught: whole-mount staining (negative staining), fixation and sectioning of cells and tissue, and shadowing of macromolecules. Each student has an opportunity to take electron micrographs of selected samples and print the pictures. Classes are arranged in the evening hours after the day's work.

Summer Seminar Series. Required of all students. All summer research trainees participate in a series of group discussions in which the students present the scope and progress of their individual research projects. The students are encouraged to ask critical questions concerning the nature of the experiments involved in the investigation. Discussions are held in the evening hours after completion of the day's work.

Special Topics. Available to advanced-level students. Students who are in their second summer of participation in the research training program conduct a literature search on specific scientific problems of mutual interest to the group. Papers on the problems are evaluated and presented orally by each member. This seminar series is designed to familiarize students with the scientific literature and to introduce the mechanics of preparing, evaluating, and delivering oral presentations.

Problems in Quantitative Biochemistry. A course designed to teach the quantitative skills essential to the everyday operation of a biomedical research laboratory. Specific attention is paid to problems such as calculations of concentrations, molecular weights, and cell titers.

Seminars. Outstanding guest speakers are invited to Oak Ridge to give formal presentations on specific topics of research and academic interest. The seminar is open to all students and faculty.

Students returning to their colleges after the first summer of training at Oak Ridge National Laboratory receive assistance and counseling from the faculty of the Graduate School of Biomedical Sciences in their efforts to select Ph.D. training programs. These students are made aware of all possible opportunities for graduate education. In addition, where possible and desirable, specific contacts for these students are initiated with Ph.D. program directors at other institutions.

Since the start of the program in 1971, a total of 67 students from 28 different institutions have received 78 man-summers of training at Oak Ridge. Of these students 60 had graduated as of June 1976. Of these 60, 23 entered medical or dental school and 18 entered graduate programs; 8 students entered the graduate program at Oak Ridge. The success of the program thus far is due to the cooperation and efforts of many organizations and individuals, including the University of Tennessee, Oak Ridge National Laboratory, and the faculty and staff of the biology division and the graduate school. We feel that we have developed a workable program that can provide students with the extra knowledge and information that will increase their chances of success in postbaccalaureate programs.

In closing, I offer a few recommendations based on my experiences with students in our summer program. First, to biology instructors at undergraduate institutions: the biology curriculum should include courses in cell and molecular biology and biochemistry, and biology majors should be urged to take mathematics through calculus, as well as a course in physical chemistry designed for biology majors. Second, to those individuals interested in organizing summer programs for minority students: I would urge you to include ample opportunities for students to practice and exercise their quantitative skills. I find that our students are very weak in their math skills and their understanding of concepts in molecular biology. The suggestions offered above will help alleviate these difficulties.

REFERENCES

Carnegie Commission on Higher Education. *From isolation to mainstream: Problems of the colleges founded for Negroes.* New York: McGraw-Hill, 1971.

Jay, James. *Negroes in science: Natural science doctorates, 1876–1969.* Detroit: Balamp Press, 1971.

National Academy of Sciences, National Board on Graduate Education. Minority group participation in graduate education, 1976.

Watson, J. *Molecular biology of the gene.* Menlo Park, Calif.: W. A. Benjamin, 1976.

Wilburn, A. Y. Careers in science and engineering for black Americans. *Science,* 1974, *184,* 1148.

Wolfe, S. *Biology of the cell.* Belmont, Calif.: Wadsworth Publishing Co., 1972.

27

The Harvard Health-Careers Summer Program

WILLIAM D. WALLACE

Accelerated recruitment by all medical schools is essential to the overall goal of increasing the proportion of minority-group physicians within our population. In order to succeed, this activity must be complemented with efforts to enlarge the national pool of minority students interested in the health sciences.

The Health Careers Summer Program (HCSP) was conceived and planned for this purpose by black medical students at Harvard University in 1968. The program's first eight-week session began on June 27, 1969, with an enrollment of 55 students. These 55 students were chosen from an applicant pool of 359. With each successive year, the number of applicants has risen. In 1975, there were 3,197 applicants for 157 places. It has become an increasingly difficult task to choose 157 students from so many applicants while assuring representation from many different ethnic groups, all of which have a compelling need for developing a large pool of highly motivated and well-prepared applicants to health-professional schools.

The HCSP endeavors to stimulate interest in medicine, dentistry, and the allied health professions among minority-group students now in colleges and to strengthen their preparation for health-professional study. The program is designed primarily for minority undergraduate students who have completed the sophomore or junior year of study. In some cases, a few students who are freshmen or who have graduated from college are also accepted into the program.

The program is based on the assumption that there are many minority students in college who are interested in the health profes-

WILLIAM D. WALLACE • Director, Summer Programs, Harvard University, Cambridge, Massachusetts.

sions but who, for a variety of reasons, apart from ability, are not likely to attain an advanced degree without academic and motivational reinforcement. An example would be the student attending a large urban university where he or she cannot or does not receive the personalized individual counseling necessary to enroll in the proper courses. In essence, this student is not only lost in the university but is also lost from the pool of potential applicants to health-professional schools. As a result, the student is not a visible candidate for graduate admissions. The HCSP seeks to increase the visibility of such students through its integrated academic and clinical programs. We feel that one way the program can affect admissions to graduate health-professional schools is by lending the prestige of Harvard University to performance that would otherwise be wrapped in obscurity. Our students are ethnic minorities and disadvantaged whites for whom HCSP intervention might mean a realistic opportunity for advanced work in the health sciences.

The participants in the HCSP are enrolled as regular students at the Harvard summer school and have access to the school's living accommodations and extracurricular activities. Financial support is provided for all students accepted into the program. An important goal of the HCSP is to provide first-rate introductory, intermediate, and advanced courses in biology, chemistry, mathematics, and physics to a large number of disadvantaged students. The importance of such courses cannot be overemphasized, since admission to medical and other health-related professional schools is dependent largely on the student's demonstrated abilities in the basic sciences and mathematics.

The primary aim of the program is, therefore, to develop a pool of qualified applicants for medical and dental schools. The HCSP is not a preparatory program for Harvard's medical or dental school. Representatives from over 50% of the nation's medical and dental schools visited Cambridge during each summer session to interview HCSP students. In cases in which the students attend small colleges, the HCSP can aid medical schools in evaluating candidates from colleges with whose graduates they have had little or no experience. For students attending urban colleges—many of whom are commuters—the HCSP provides an opportunity to devote a full summer to premedical study free from interruption and financial worry.

A secondary aim of the program is to give the students exposure to hospitals, laboratories, and community health problems. It is hoped that this experience will help the students to assess their interests in the health professions realistically.

The program consists of six main activities: (1) course work; (2) small-group tutorials; (3) clinical tutorials; (4) group and individual counseling sessions; (5) professional-school interviewing; and (6) study-skill workshops and MCAT reviews. The Faculty of Arts and Sciences provides the course instruction, while the Faculty of Medicine provides the clinical tutorial.

After taking a placement examination and in consultation with the staff, each student is enrolled in a regular course at the Harvard summer school and proceeds at the same pace as other regularly enrolled Harvard and Radcliffe students. Consequently, none of the courses taken by HCSP students is remedial. Courses in chemistry, biology, mathematics, or physics are open to HCSP students according to their interests, academic backgrounds, and needs.

On the basis of the experiences of the first two sessions, additional courses have been made available. In 1970, "Selected Topics in Cellular Biology" was initiated. Although open to all summer-school students, the course was developed with HCSP participants in mind and was taught by the Medical School faculty. Because this format proved especially popular with HCSP students and Harvard–Radcliffe students, a similar course in nutrition and health sciences was initiated in 1974.

The summer-school courses are supplemented by academic tutorials with a ratio of one tutor to three or four students. The tutorial staff is recruited from the teaching fellows in the science area of the Faculty of Arts and Sciences and from advanced students in the Medical School, the Dental School, and the School of Public Health. A majority of the tutors come from the ethnic minorities represented in the HCSP, and some are former HCSP participants. These sessions are designed to stimulate maximum individual participation. During the eight-week session, an academic tutor comes to know his students personally and can identify and work with their individual areas of strengths and weaknesses.

The clinical tutorials involve a minimum of two days per week at one of the 11 health-care sites affiliated with the Harvard Medical School. Almost every conceivable aspect of the health-care experience is covered by this group of facilities. Students are assigned to individual preceptors and receive intensive exposure to the medical experience in a way otherwise impossible. We consider the clinical experience and the resources available a unique feature of our program. We also create additional workshops and seminars for HCSP students. In 1975, there were such workshops ranging from "How to Study for the MCAT" to "Minority Women in Medicine."

One of the most important secondary activities offered by the HCSP is premedical and predental counseling. Many of the administrators and faculty involved in the HCSP have been distressed by the poor preprofessional counseling services available to HCSP students at their home schools. Consequently, a substantial counseling program was established.

Another equally important part of the HCSP is the medical- and dental-school recruiting program. A letter is sent to the admissions office of each medical and dental school in the United States before the summer, inviting a representative to meet with HCSP students in Cambridge. Schedules are set up so that the students may have individual interviews with as many representatives as they wish. The final measure of the effectiveness of the program is the successful entry of its students into health-professional schools. Of course, it is difficult to determine the extent to which the HCSP contributes to this accomplishment. However, we have data from the Association of American Medical Colleges that shows us that over 70% of the HCSP students have been accepted into a health-professional school.

In an evaluation report, sponsored by the National Center for Health Sciences Research and Development, based on the initial two years of HCSP, it was established that programs such as HCSP could be organized in other educational institutions if the following conditions exist:

1. The university or college must have in existence a large, active summer school with strong science courses.
2. Medical and dental schools and affiliated hospitals and clinics must be located near the university.
3. Students enrolled in the programs must function as a part of the regular summer-school student body.
4. An adequate number of minority medical/dental students must be available to serve as role models.

Regional availability of programs such as the HCSP may also be possible. Such offerings could reduce the operational costs by lowering the funding required for transportation, tuition, and room and board.

Since its inception, the HCSP has gradually included more ethnic groups. Each year, our admissions committee wrestles with the problem of choosing criteria for an equitable distribution of available places among minority groups. Should the distribution be based on the needs of a group? American Indians and Puerto Ricans say yes. Blacks and Chicanos say no. Should the distribution be based on national populations? Blacks and Chicanos say yes. American Indians and Puerto Ricans say no. Clearly, there are many difficulties in the operation of a multiethnic program. However, these are far outweighed by the advantages of providing for cross-cultural interaction.

The HCSP has grown in stature and in numbers since it was initiated in 1969. It is important to keep programs such as the HCSP alive in these times of economic insecurity and increasing hardships for the nation's poor. Opportunities to participate in programs such as the HCSP can have a positive impact on the career aspirations of underprivileged students. This country's health needs will not be fully met until it can provide training for health professionals from every ethnic and socioeconomic group.

28

Postbaccalaureate Premedical Programs for Minority Students

JEWEL PLUMMER COBB

The delivery of adequate health care is one of the most challenging domestic problems in the United States today. Its solution by way of national health insurance or the extension of current Medicare and Medicaid programs, plus rapid implementation of biomedical-research findings, is only one part of the story. The more serious challenge that has not yet been approached is the training of manpower to serve the poor and marginal-income citizens of our burgeoning urban populations and our lonely and isolated rural towns. A significant human resource for providing the pool for the medical workers needed is the 21% minority population in our nation.

I describe here a plan that enlarges the pool of minority applicants for the health professions in a very short time with minimal cost. As an efficient and economically feasible program, it provides the student with the necessary academic preparation to be an eligible medical-school applicant.

The Association of American Medical Colleges recognizes the physician shortage in the United States and points with concern to the woefully neglected medical care for oppressed minority and disadvantaged populations. The accepted minimum standard of safety is 1 physician to 1,500 of population, and the national average is about 1 to 750. But in the black ghettoes of America, the proportion of black physicians is 1 to over 4,000 blacks, for white physicians rarely practice in the ghetto community. According to the 1972 census, there were only 5,478 black physicians of the 320,903 total of active physicians. There-

JEWEL PLUMMER COBB • Dean, Douglass College, Rutgers University, New Brunswick, New Jersey.

fore, given the Census Bureau's estimate of 23.5 million blacks, there is 1 physician for every 4,298 blacks (Thompson, 1974). The plight of the American Indian, the Mexican-American, and the Puerto Rican population is even worse.

The Association of American Medical Colleges has declared that there should be a national effort to increase the number of students traditionally underrepresented in medicine. If this is to be accomplished, the organization suggested that the minority enrollment increase to a minimum of 12% by 1975. This minimum was not achieved, however, according to the recent study sponsored by the Josiah Macy, Jr. Foundation; the increase was only 8.2% (Maxine Bleich, this volume). In 1974, the enrollment figure was 6.3% for black Americans, 0.3% for American Indians, 1.2% for Mexican-Americans, and 0.3% for mainland Puerto Ricans (total 8.1%). Clearly, a major problem is that of providing an adequate pool of eligible students by focusing efforts on programs in earlier training, recruitment, and retention.

An important untapped pool of minority resources is the young, mature, adult college graduate with demonstrated learning skills who wants to redirect his or her career to medicine or dentistry. This postbaccalaureate candidate has only recently been given encouragement even to consider a professional health career, since enrollment figures indicate that as late as 1967, only around 180 students from all historically black colleges were entering medical schools. In 1968, the numbers increased; yet 31 medical schools did not enroll a single black student. Currently, there are black and other minority students in most medical schools.

In 1966, the Josiah Macy, Jr. Foundation gave a grant to Haverford College in Haverford, Pennsylvania to expand its postbaccalaureate-fellowship program to include students interested in careers in medicine. The program provided for supplementary study for disadvantaged college graduates who could not compete with other college graduates for admission to graduate or professional schools.

A few years later, we at Connecticut College were stimulated by the Parham Smart Study prepared for the Medical Care and Education Foundation. This study revealed a marked discrepancy between the total minority-group population in New England and the percentage of minority-group physicians. While blacks comprise 2.3% of the total population in the six northeastern states, they provide only 0.5% of the region's doctors. We therefore began to plan a postbaccalaureate premedical/predental program for the mature nonscience graduate who wanted to redirect his career goals. In 1971 and 1972, the Grant Foundation, the Sloan Foundation, and the Van Ameringen Foundation awarded grants to initiate a program for promising minority students interested in medical- or dental-career courses. The program provided for a year of academic instruction in premedical sciences and in its first year attracted a group of competent, well-motivated, and enthusiastic students. In view of the success of this program during its first year and the promise for the future, an additional two years of support was provided by the Grant Foundation.

Without private foundation support, the program would never have been possible. The postbaccalaureate student is not eligible for the usual undergraduate scholarship; neither is he eligible for medical-school financial support. Without support from a program like ours, such promising students could not finance the redevelopment and redirection needed for a year of full-time study. Program money from government sources should be made available for the postbaccalaureate student who wishes to redirect his career toward the health professions.

Students are selected on the basis of their undergraduate performance, their potential for academic achievement, and their maturity for adaption to the new college and community environment. Prior to entering the program, the student must have one year of inorganic chemistry with good grades.

The selection committee consists of medical faculty from Albert Einstein, Yale, and the University of Connecticut, plus a zoology professor, a student in the program, and the chief administrator of the program. This spring the committee selected 10 students from a group of 129 applicants.

The following supporting services are provided to the students: (1) advice on medical-school application procedures; (2) secretarial services related to medical-school applications; (3) weekly tutoring assistance on a volunteer basis by scientists at the nearby Pfizer Pharmaceutical Company; (4) psychological counseling and care; (5) academic counseling by the director and by the faculty teaching courses; and (6) visits to medical schools, including tours and discussions with deans, students, and admissions officers.

During the academic year, at the college, each student receives a small stipend plus tuition, books, and prepaid student-comprehensive fee. They are required to take the Medical College Admission Test (MCAT) in September and again in May, at the end of their academic tenure. During the year, they practice taking mock MCAT tests under simulated testing conditions. They are counseled on their achievement, and gaps in their learning are pointed out so that they can study the needed material.

The students apply to professional schools in the fall and are interviewed during the academic year while in the program. Some are accepted in the late winter and some are placed on hold lists, pending receipt of the second-semester grades and the May MCAT tests.

Students enrolled in the program had majored in various fields as undergraduates, including French, economics, English, black studies, or psychology. All had graduated with very good cumulative grade-point averages. A few had taken math or science courses as undergraduates. The program is now (1976–1977) in its fifth year. A total of 29 students have completed the program, 25 of whom are now in medical school.

The program has proved to be cost effective in that mature, bright minority students have been prepared for medical or dental school in one year within a budget of less than $4,500 per student (including costs and fringe benefits). Our success has been such that a number of admissions offices now look forward to interviewing our students. The acceptance to medical/dental schools has been high. In two years, we will have the first full-fledged physician from the program.

REFERENCES

Medical Care and Education Foundation. The Parham Smart Study: An internal document of an informal consortium of New England medical schools to assess the state of the medical schools in the region. (Unpublished.)

Thompson, T. Curbing the black physician manpower shortage. *Journal of Medical Education,* 1974, *49,* 944–949.

VII. Special Training Programs for Minority Students in Science: Precollege Level

29

Training at the Collegiate and Precollegiate Interface: STRIKE as an Example

J. N. GAYLES

From 1974 to 1976, Morehouse College operated a program designed to increase the pool of minority students in a pathway leading to health careers. This program was a high-school program operated from the base of eight predominantly black high schools in the city of Atlanta.

Morehouse College, a men's college in Atlanta, Georgia, has a distinguished record in science and premedical education. This 109-year-old institution has more alumni holding M.D., D.D.S., and Ph.D. degrees than any other predominantly black college.

Given the national ratio of 1 black physician to 4,510 black people, Morehouse College has recognized for decades the need to go beyond what might be considered an average institutional emphasis on the area of medicine. More than a tenth of the nation's black health professionals graduated from Morehouse. Of over 5,000 living alumni, 450 are physicians or dentists, and they represent 6% of all black physicians and dentists in this country. Each year, the college places 12–20 students in medical and dental schools. Of the total student majors in science, at least 95% go on to graduate or professional schools.

Morehouse College also has an outstanding record in the production of basic scientists who hold the Ph.D. degree. To date, there are 67 known recipients of the Ph.D. degree in the basic sciences who received their bachelor's degree from Morehouse College. Particularly in the field of chemistry, the Morehouse record is impressive. In a 1973 article based on a study conducted by the U.S. Office of Education,

J. N. GAYLES • Professor of Chemistry, Morehouse College, Atlanta, Georgia. This project was supported by PHS-DHEW-HRA-OHRO 04-D-001198-01-0.

which appeared in *Chemical and Engineering News* (1973), it was stated that less than 2% of all American chemists are black. This 2% constitutes approximately 225–250 black Ph.D.'s in chemistry. This figure should be compared with 46,000 Ph.D.'s awarded in the chemical sciences over the past 40 years. Of the 225–250 black Ph.D.'s in chemistry, 34 are Morehouse graduates. Morehouse has well-established departments in all of the basic science areas. Also, the college has most recently become involved in developing a two-year school of medicine.

It was into this atmosphere of historic interest in health care and education in the sciences that approximately 60 high-school students were drawn for participation in the academic reinforcement and motivational program.

The data cited above clearly indicate that there is a desperate need for an increased number of black health-care professionals. Particularly in the field of medicine, the shortage of black professionals is acute. Morehouse has sought to address the shortage of black professionals in medicine in two major ways. As previously stated, efforts are currently under way to establish a two-year school of medicine at Morehouse College. Also, it was the belief of the staff and faculty of the college that the pool of students seeking preprofessional training for medicine could be enlarged, and improved with respect to preparation, by special attention to the educational goals of students as early as the ninth grade in high school. With this in mind, an effort was made to supplement the ongoing educational activities in selected Atlanta inner-city high schools in order to prepare students better for careers in health. An attempt was made throughout the high-school program to develop mechanisms for the early recognition of ability and potential. With these data in hand, we attempted to move toward minimizing the negative effects of inadequate high-school preparation on the college training process. Thus, the high-school program STRIKE was developed to provide a vehicle for college professors, medical students, college students, and medical-school faculty to begin working with a pool of students who, we felt, had the potential of becoming excellent health-care professionals following their training process.

STRIKE is an acronym developed by the students, which stands for "Special Tactics Research in Keeping Everybody." This acronym attempts to encapsulate the feeling of all of the people working with the program. That feeling was "each one teach one"; in other words, a team approach was developed to simultaneously maximize the number of people who start out in quest of a career in health and minimize the number of students who fall by the wayside.

Following extended meetings with a planning group composed of high-school counselors, high-school science teachers, medical students, college faculty, and college students, an attempt was made to develop a sequence of activities that would serve the two basic purposes of the program. These purposes were academic reinforcement and motivation. On the one hand, there was a desire to strengthen the preprofessional preparation of students for college work by attacking those academic areas in which locally administered tests indicated that the students had general weaknesses, primarily the areas associated with competence in quantitative skills. In particular, the areas of precalculus mathematics and chemistry were noted to be areas of serious deficiency on the part of students. Weaknesses in biology were not as severe, and perhaps because of this fact, there was a high level of enthusiasm for using biology as a *motivating* mechanism in order to keep students interested in health while developing the requisite quantitative skills.

The planning team evolved a sequence of activities for the students that is summarized in Tables 1 and 2 and in the Appendix (see p. 251).

Table 1 presents clearly the emphasis in the programs on quantitative skills as well as the more life-science–oriented activity (a more complete listing of these data are available upon request). Table 2

Table 1. Project STRIKE, Academic Core Schedule II, Morehouse College

Date	Mathematics/Biology	Chemistry
Week 5, Feb. 14	Logarithms	Applied Problems in Chemistry
Week 6, Feb. 21	Significant Figures and Algebraic Equations	Chemical Bonding I
Week 7, Feb. 28	Review for test	Review for test
Week 8, March 6	Midterm test	Midterm test
Week 9, March 13	*Biology* Anatomy of the Cell	Chemical Bonding II
Week 10, March 20	The Central Nervous System	Redox Reactions
Week 11, March 27	Genetics and Man	Gases, Liquids, Solids
Week 12, April 3	Drugs and Man	Nuclear Chemistry
Week 13, April 10	Parents and students' career day—manual for parents	
Week 14, April 17	Program OUT—Easter break	
Week 15, April 24	Review for test	
Week 16, May 1	Testing	
Week 17, May 8	Culminating-activities awards	

Table 2. Project STRIKE, Morehouse College

Date	Summary of Health and Medical Conference Events
February 14	A beef heart will be dissected. Comparison will be made to anatomical structures of the human heart. A film, *Hallah-SOS,* will be shown. This movie will emphasize the recruitment of blacks for the health professions.
February 21	Dr. James Story, professor of anatomy and biology at Morehouse, will give a stimulating lecture on aspects of the human reproductive system.
February 28	Mrs. Jean Floyd, Assistant Director of the Outreach Family Planning at Grady Hospital, will lecture on various aspects of family planning.
March 6	Dr. Audrey F. Manley, a pediatrician at the Emory University School of Medicine, will lecture on minorities and women, the "right" to medical care as opposed to the "privilege."
March 13	A laboratory session will be held using blood-typing kits. Students will be taught the importance of hematology and how to determine their blood type. A film, *A Matter of Opportunity,* will be shown. This film relates to minority health manpower shortages.
March 20	The medical-student staff will present a group lecture on cancer entitled, "Cancer Reversal to the Embryological State."
March 27	The medical students will present a group lecture and demonstration on mental health and its epidemiological aspects in the black community.
April 3	Dr. Louis Sullivan, Dean of the School of Medicine at Morehouse College, will present a lecture on blood diseases with particular reference to "Sickle-Cell Anemia—Facts and Fiction."
April 10	Parents and Students' Career Day will consist of various informative seminars given by invited guests from the allied health professions. A "STRIKE Parents' Manual" will be distributed and discussed.
April 17	Easter break
April 24	Student Day—Each high-school *captain* will give a 15-minute presentation on precollege science activities at his or her school.
May 1	College "buddy" day—College students and high-school students will be paired. Aspects of various college activities will be explained: cost, activities, subjects, etc. Classes will be attended by all high-school students.
May 8	Culminating activities and award presentation—Morehouse awards will also be presented at Honors Day at each participating high school.

presents a summary of the health- and medical-conference events that took place during the latter portion of the program. Table 2 presents typical events that are reflective of the more motivational and informational aspects of the program.

The Appendix shows the form and substance of one of several planning documents generated by the planning group. This synopsis of some of the implementation aspects of the program demonstrates the ability of our model to learn from past errors and institute self-correcting measures.

A sequence of examinations was presented at the commencement of the program in order to establish basic norms of student performance in each area. This set of examinations enabled us to divide the students into smaller groups so that they could be taught academic skills based on the level of their attainment. This approach proved to be very satisfactory for the instructors in the program as well as for the students. It should be stressed that in terms of an instructional setting, it is probably more useful to have a fairly homogeneous group of students in terms of academic level than to attempt to deal with the widely varied problems of a heterogeneous group.

The testing program continued on a periodic basis so that we could continue to sample the progress of the students and make them aware of and more comfortable with examinations of the type normally associated with "standardized" tests. Finally, examinations were given at the end of the program in order to gauge progress and to provide a meaningful basis for program-exit counseling.

All activities for our program were conducted from 9:30 A.M. to approximately 12:30 P.M. on Saturday mornings. This schedule aided in the selection of students who were serious about obtaining the best preparation for a professional career. The students who remained throughout the course of the program (approximately 80%) were highly motivated and were very serious about their career intentions. And as a result of the program, they were better prepared academically to proceed to the next level of their high-school training or to go on to collegiate training for a career in medicine.

A number of other small events took place during the course of the program that are worthy of mention. A number of prominent professional guest lecturers were invited to speak to the students in the interests of conveying useful scientific information and to provide students with positive role-models of individuals who have proceeded through the process they were facing. It was especially important to provide the students with examples of minority individuals who had succeeded in developing skills to the point where they were successful professionals. The image of the successful minority professional provided the students with a vital and frequently utilized reference point, which was too often absent from the home or the community environment.

A college "buddy" day was arranged for each of the students, on which a college student assumed responsibility for a high-school student. The high-school student attended the buddy's college classes, toured college facilities, and met and spoke with college professors in order to develop a feeling for the collegiate education process—its difficulties and its positive features as well.

Tours of local health-service institutions were also arranged for the students. These were limited in number because of the fairly large number of students involved in the program. Even dividing the group into small units did not prove to be of particular value for the effective transfer of information. Our results suggest that tours are not easily developed into an effective mechanism for academic reinforcement or for motivation.

The planning committee decided to work with four large, predominantly black high schools in the city of Atlanta. It was the initial intention of this group to involve 10 students from each of these high schools, with each school having a student team "captain" as a local coordinator. Because of the high level of interest expressed in the program, we began with a group of approximately 60 students, and because of the normal attrition process, we ended with a group of approximately 50 students. Additional high schools were involved as word about the program spread. As we proceeded, we evolved from our effort with the four initial high schools to involvement with eight major public high schools in Atlanta. A planning group of approximately 10 individuals worked intensively to put together the course and the curriculum activities described above. Two program "graduates" were later added to the planning group.

A successful initial effort was made to develop an inventory instrument to use for the purpose of evaluating potential students. However, for the most part, the students were self-selected; that is, almost all students who expressed an interest in participating in the program were allowed to attend for a period in order that each might be evaluated. The students also had an opportunity to evaluate our program. Following approximately three to five initial meetings, a natural attrition took place, and for the most part, the students who we felt, for a variety of reasons, were the most "difficult cases" chose not to return or to continue participating in the program.

The students involved in the program were in the 9th, 10th, 11th, and 12th grades at one of the high schools associated with the program. Our initial concept was to work with students in the 10th, 11th, and 12th grades. However, in view of the high level of interest of some 9th-grade students, we began working with them also. It is worthy of note that many of the best students in the program, academically, were from the 9th grade. Our experience is that the deficits in performance become more substantial the further the student proceeds through the public educational system. Thus, we felt it was most reasonable to work with students in the lowest grade possible.

The teaching staff of the program was selected largely from the initial planning group. In addition, because of federal support, we were able to add several college students and medical students who worked on a weekly basis with the high-school students on Saturday mornings.

Because we are situated in the city of Atlanta, we were able to draw upon the resources of the active medical school in the city of Atlanta, Emory University College of Medicine. Several of the black students at the Emory University College of Medicine chose to work with us in a very enthusiastic manner as we attempted to implement our high-school program. These medical and college students were involved in planning, in management, and in detailed program implementation. Because we had excellent medical and college students, the program proved to be rather effective. It is fair to say that without them, the program would not have succeeded. We also involved college faculty, high-school counselors, and high-school science teachers.

A word about our high-school counselors and science teachers would probably be worthwhile. Particularly from the perspective of college faculty, the high-school counselors and high-school science teachers are a much-maligned group. Our experience in this regard was rather encouraging. While we were rather selective in the counselors and science teachers with whom we became involved, we were encouraged by the number and the quality of high-school science counselors and science teachers with whom we worked throughout the course of the program. They were imaginative and resourceful, providing us with very keen insight as to the best mechanisms for achieving goals with high-school students because of their familiarity with nuances in the preferences of the high-school students. They also were able to give us perspectives that related to jobs that high-school students have, obligations to Saturday afternoon band or football games, and things of that sort. They were also involved at the very substantive level of course and curriculum implementation.

Over the past two years, approximately 80 students have been involved in the STRIKE program on a continuous basis. Survey instruments that have been administered on a regular basis indicate that these students have the opinion that the program has benefited them in an academic as well as a motivational sense. Of the 80 students who have been involved, approximately 23 have "graduated" from the program. Our records indicate that all 23 of these students are currently enrolled in a college in the United States. Three-quarters of the students have enrolled in one of the colleges in the Atlanta University Center (Clark College, Morehouse College, Morris Brown College, and Spelman College). About 5 of these students have enrolled at Morehouse College. All of the 5 STRIKE enrollees at Morehouse are performing in an above-average manner. This performance should be viewed in the light of comparative overall weak initial performance in our high-school program. In terms of placement of students in college for preprofessional training, we feel the program has been a success.

There are other less tangible evaluations that lead us to believe that the program is quite positive in its overall impact. For example, we have observed STRIKE students who began with us as high-school students in the 11th grade who are now college students attempting to deal with the day-to-day strife of a college education in the sciences and at the same time working with us on weekends with high-school students, still participating in the STRIKE program. The STRIKE graduates are particularly useful as interpreters between the high-school students and the instructional staff. Also, apart from the rather minimal monetary concerns, we feel that the academic elements of the program have particularly benefited our college students and medical students who have participated in the program. That is, as they have had to explain basic scientific principles, they have had to learn and relearn a great deal themselves. In fact, the medical students have particularly valued the human contact that they have gained working with high-school students on Saturday mornings. Most of the medical students who have been involved in the program have been very personable and human-oriented. They apparently value a life experience at this stage in their training, one that takes them away from intensive medical classroom work and involves them with students in their formative stages.

Finally, apart from demonstrable improvements in the areas of academic attainment and quantitative skills, increased clarity of purpose, and goal orientation, there has been a gratifying sense of overall movement on the part of students. Even in the brief span of operation of the program, we have seen our high-school students become college students interested in and working with us to improve the opportunities of the high-school students who are still coming along. We have seen some of the college students who worked with us continue to medical school with increased levels of academic ability and increased sensitivity to the needs of the people in their communities. We have also seen medical students who were thinking of certain speciality areas who are now in postgraduate training programs that involve the areas most closely allied with primary care. We have seen, in one case, a postgraduate medical student who was involved with our program move into the practice of medicine in a setting that provides for primary care. Thus, the program has shown us that it is possible to make a difference even in a small way for students who have latent ability and interest.

Through contact with the college students, the medical students, the college faculty, the high-school science teachers, the high-school counselors, and the parents of the students themselves, we have witnessed an enormous investment of time, intelligence, caring, and, to a lesser degree, money. While we have not "kept" everybody, we have "kept" most. We feel that the rewards have more than justified our investments.

Appendix

Planning Group Document—I

Since the time allotted for the Saturday program last year (two hours in 1974–1975) seemed somewhat inadequate, the program will be held for three hours each week. With modifications where necessary, the schedule will follow the following outline:

Total time	3 hours/9:30–12:30
Math	45 minutes/9:30–10:15
Clinical correlation	45 minutes/10:25–11:10
Chemistry	45 minutes/11:20–12:05
Open discussion and announcements	25 minutes/12:05–12:30

A rigid and thorough outline will be followed for the core presentations in mathematics and chemistry, beginning with basic concepts that may have previously been covered in high school (probably only by a few) with work toward a designated goal or learning objective for each module. Material should be so outlined that a fixed amount is covered in one session, so that the students as well as the instructors have the satisfaction of completing a prior goal each week.

The clinical-correlation block is designed as the catchall for the nonacademic portion of the program. This time should be used for films, guest speakers, student presentations of projects, and medical correlates presented by the staff. It must be emphasized that this time must not encroach on the main academic-core presentations. This portion of the program will require at least as much planning as the core academic portion. It should be arranged so that the students do not get a barrage of films followed by a lengthy series of lectures, etc.

The open-discussion period can be used for making announcements as well as providing the students with an opportunity to pursue some point of interest that might have been brought up in either the academic presentations or the clinical correlations. This time also would be good for doing things like discussing blood pressure and taking blood pressures, etc. It is important to plan activities that involve physical movement and the use of tactile skills in this portion of the program.

One staff member should be designated as being responsible for procuring the needed audiovisual equipment, seeing that the appropriate rooms are open, etc. Another staff member should be responsible for maintaining files on the students. This file should contain the student's application (survey inventory), a picture of the student, and a critique of the student by his high-school counselor. Subjective and quantitative evaluations should then be added to this folder periodically by the staff members presenting the core curriculum.

Institutions (e.g., Southwest Community Hospital) might be more amenable to having smaller groups tour their facilities; therefore, it will be better to schedule several trips throughout the year so that five or six students at a time will go on the tour. This plan will increase the probability of their gaining substance from the experience. However, it must be emphasized that these are only adjunct activities of the program and should not encroach on the core presentations. This means scheduling them after 12:30.

REFERENCE

Chemical and Engineering News. Less than 2% of chemists are black. January 8, 1973, p. 25.

30

Motivating Upper-Elementary-Level Mexican-American Students toward Science Careers

ROBERT A. WARREN

A great number of programs have been designed to bring minority-group members into careers in science. They are helpful; however, they serve only as stopgap measures and are not designed to develop a continuous, ongoing pattern of career orientation toward science. What should be developed, therefore, are projects specifically directed toward the development of a continuous, year-after-year stimulation of interest in science and careers in science for all students regardless of race or ethnic background. If students are to be developed who are *really* interested in science and science careers, such individuals should be "turned on" to science at an early age, by interested, informed, and motivated teachers. These would be teachers at the elementary-school level.

Texas A. & I. University in Kingsville, Texas, through grant assistance and encouragement from the National Science Foundation, has embarked on a project to provide stimulating science experiences for elementary students. Because of its geographic setting in south Texas, Texas A. & I. University focused its attention on the Mexican-American elementary student. Mexican-American students are bilingual, with English being the second language, and comprise over 70% of the school population in this vast, mostly rural section of Texas. Very few Mexican-American students enter into any kind of science-related careers.

In the development of the project at Texas A. & I., several basic

ROBERT A. WARREN • Associate Professor of Education, Texas A. & I. University of Kingsville, Kingsville, Texas.

assumptions were made about a student's motivation to enter a science career. First, we assumed that one must be confident of doing well in science and must have experienced success with science at an early age in school. Second, one must have an innate desire to learn more about science, a desire to explore, to question, to seek information about the physical and living world within one's own life. Third, one needs to have information about and realistic impressions of the total spectrum of career opportunities and employment trends in science and science-related fields.

Using these assumptions, Texas A. & I. developed a program to help science teachers of upper-level-elementary, bilingual Mexican-American students motivate their pupils toward considering career opportunities in science. It was recognized that these teachers first needed to have a solid foundation in the basic sciences. They also needed to develop skills and strategies that would be effective in reaching and teaching bilingual students, as well as an awareness of job opportunities in science both at the national level and within the geographic area in which their students live. These teachers also needed to develop the skills necessary for counseling and guiding the Mexican-American student toward future course work and careers in science.

To provide teachers with these skills, Texas A. & I. initiated a one-year, five-phase training and assessment program. The first phase of the program was a six-week workshop for 18 upper-level elementary-school teachers. The workshop was held during the summer and consisted of full-day sessions. All teachers received graduate credit for the workshop, and a stipend was also provided. Of the 18 participants, 10 were bilingual. The main emphases of the summer program were to develop strategies for teaching bilingual students and to gain some familiarity with local industrial organizations that employed scientists or engineers. The latter point was achieved by the sponsoring of field trips to such sites as the Brook Aerospace Medical Center, the Kingsville Naval Air Station, PPG Industries, Sun Company Refineries, and the Corpus Christi Weather Bureau. These trips proved to be a great success. The teachers were able to learn about the jobs, the salaries, the job opportunities, and the training requirements for various science-oriented professions. At the conclusion of the visits, the teachers were required to develop appropriate science-career information files for their particular grade level.

To help develop teaching strategies for the participants, the program employed specialists in bilingual education. These specialists used the basic language-arts skills of listening, speaking, reading, and writing. A major emphasis was placed on the development of bilingual slide–tape programs that could be utilized in large groups or provide individualized instruction for students with language difficulties. Another area of concentration was that of writing bilingual scripts to help students develop an adequate scientific vocabulary and enhance pronunciation skills.

At the beginning of the school year in September, teachers who had participated in the summer program were paired with teachers who had not received the training workshop. Teachers were matched according to grade level, the approximate percentage of Mexican-American students in their classes, and the bilingual abilities of the teachers. Students of teachers in the experimental and control groups were given science achievement tests. They were also administered the Kuder Form-E Interest Inventory or the SRA "What I Like to Do" inventory to determine their interests in science careers.

Throughout the following school year, teachers in the experimental group received additional training by attending Saturday workshops, totaling 45 hours of instructional contact. Part of the Saturday sessions focused on obtaining indepth exposure to information in the different science disciplines. A good deal of emphasis was placed on laboratory material that could be used for display by elementary teachers. Each meeting focused on a different science discipline, featuring a guest instructor from the appropriate university department. Another regular feature of the Saturday meetings was the sharing of practical classroom ideas for teaching science. The teachers made suggestions for science-career orientation and bilingual instruction using their own experiences as a guide. This format provided many varied suggestions that could improve science instruction.

At the end of the school year, statistical analysis of the results of the science achievement tests and the science-career interest inventories of the students proved disappointing. We did not observe a significant difference between the test scores of the experimental and the control groups.

Although statistically there were no significant changes in the experimental and control groups, there were many other indicators of the project's success. First, many of the control-group teachers sought and initiated new approaches in science instruction and cooperated with the experimental group in science instruction. They participated in field trips, science fairs, and film showings and used the bilingual slide–tape presentations developed in the summer program. Although this relationship invalidated any statistical inferences, it showed rather dramatically how teachers in the project had learned and developed as a direct result of their participation and how they were sought out by their peers for assistance. Discussions with school principals indicated that the teachers who participated in the project had given a certain direction to their teaching programs.

Another factor that stymied the statistical analysis of the project was the finding that only 28% of the experimental group of students and 26% of the control group produced test scores that had a sufficient amount of internal validity. Because of the poor language skills of the bilingual Mexican-American students, approximately three-fourths of all students in the control and experimental groups could not provide valid information, using standardized testing, that could attest to the success of the project. This experience shows a dramatic need to develop valid testing measures for the Mexican-American student.

In conclusion, it is the opinion of the project staff that within a few months, a great deal can be done to assist teachers of elementary-school bilingual Mexican-American students to enhance their science skills. Such a program can, by simple diffusion through peers, help to provide a continuous, ongoing pattern of career education for these students for years to come.

31

Ethnoscience: An Educational Concept

CARL HIME

As a concept in education, ethnoscience connotes the use of a student's home-, community-, or culture-centered experiences to teach concepts or relationships in science.

The original intent of the ethnoscience program was two-faceted. First, there was an effort to interest Navajo students in science and to develop their knowledge of the subject and, second, to develop an interest in science careers. These two purposes, though closely related, address themselves to two different kinds of students. One type of student is "turned off" or, at best, neutral to science; the other is curious and potentially excitable about some specific area of science.

In developing the idea of using culture-centered instructional materials in a science classroom, we found it necessary to enlarge two commonly used terms. The terms are *culture* and *community*. *Culture*, for American Indian education, has generally implied the American Indian experience as it existed 100 or 400 years ago. This connotation of the word *culture* is not realistic or useful. Navajo students have experiences that are unique to them as American Indians and yet quite different from those of their forefathers. Although Navajo students grow up in the Indian traditions, their parents expect them to adopt the best from both their own and the general American culture. As Navajo students and many of their parents live in a highly transcultural environment, the term *culture* must therefore connote the entire spectrum of life experiences instead of a set of uniformly identifiable experiences. The Navajo people are characterized by a diversity of experiences. Their culture, like the general American culture, has both traditional and

CARL HIME • Director, Office of Academic Services, Many Farms High School, Bureau of Indian Affairs, Many Farms, Arizona.

contemporary elements. Navajo culture is not only what it was 200 years ago but also what the Navajo students experience today.

The second term is *community*. Here again we enlarged the idea to include not only Chinle, Arizona, or Many Farms, Arizona, with their people and setting, but more generally the entire geographic area in which our students live, work, or visit.

Students might herd sheep at Lukachukai, Arizona, but they probably buy their groceries in Farmington, New Mexico, 100 miles away, and then spend their summers in Los Angeles, Denver, or Phoenix with cousins or parents. The fact is that many of our students frequently travel over much of the Southwest.

In addition to the above definitions of *culture* and *community*, and as background, I would like to comment on the philosophy that served as a basis for the development of our particular program. The philosophy is implied in the definitions; however, briefly stated, it is: cultures in the Americas tend to interface, and elements of interfacing cultures hybridize. The recognition and the use of this phenomenon are essential ingredients in the evolving transcultural development of minority populations. For maintenance of cultural health in both the cultural minority population and the larger pluralistic American culture, mechanisms are needed to shape the hybridization process to the needs, interests, and desires of the minority people. For this to occur, one must have mutual understanding of a multicultural model of people–environment, people–technology, and people–people interactions. In developing the ethnoscience philosophy, we have limited ourselves to science education because of the small number of Navajo Indians with training in science and because modern technology has and will continue to have a sizable impact on the social and technological patterns of the Navajo people.

Before describing the approach itself, I would like to comment on some of the questions that can be asked to determine students' interest in and motivation toward science careers. First, a student should be aware of the science-career options that are available to him. Career opportunities exist in health, engineering, and physics, for example, and should be pursued. Second, we encourage a positive emotional association with science and health professionals (e.g., doctors are generally good people; scientists are not all evil, crazy people). Third, a student should be able to associate himself rationally with a career in science. He should be able to see himself as a doctor, or a scientist, or a laboratory technician. Next, the student should be involved in some type of self-assessment. For this, he has to ask some questions of himself. Is he willing to discipline himself to study? Does she use mathematics and English as useful tools? Does he enjoy discovering why or how things work? Does she faint in a hospital (as a former student of mine did on her first trip to a hospital), just after she had decided that she wanted to be a hospital nurse? If the above points create positive impressions, then a student is encouraged to pursue plans of study that will lead to a career in science. For this, the student needs access to a college program in which he or she can succeed and subsequently realize the career goal. These last two steps are outside the scope of this paper, but suffice it to say, sometimes these are the most frustrating steps for a secondary-school teacher who is several years removed from the student's college graduation.

One difficulty with the evaluation given above is the wide range of reasons or motives for students entering a science career. This diversity leads to difficulty in the designing of materials that will reach many or any students in a particular class. We have found, however, that five areas of motivation are frequently mentioned by our students. These are (1) models (e.g., a Navajo medical doctor); (2) helping his or her own people; (3) potential earnings; (4) curiosity about how things work; and, finally, (5) a motivation described as a student's feeling that he or she wants to do it for some "gut" reason. Including these areas in the evaluation has led to an expanded range of activities as partially described later. The following is a description of the ethnoscience program developed for our Navajo students.

THE SETTING

Located approximately 250 miles from the nearest city of any size (Albuquerque, New Mexico) and 100 miles from the nearest town (Gallup, New Mexico) the school where ethnoscience has been used most is in Many Farms, Arizona, at the center of the Navajo Nation. The students are Navajo youngsters attending Many Farms High School of the Bureau of Indian Affairs. For these students, English is the second language; much of the school culture is foreign to their home communities.

THE PROGRAM

Ethnoscience started in 1966 at Wingate High School when Navajo custodians and Navajo students in my classes asked questions about the relationship between science and the Navajo culture. The students did some beautiful classroom projects, which were used for the science fair and in turn became instructional materials for the following year's biology students. The projects involved the use of Navajo vegetative dyes in pH indicators and as bacterial-growth inhibitors. By the end of the second year, the list of projects had grown to 27 different topics ranging from blood-type comparisons to soil analysis, and by 1971 at Many Farms High School, the list was considerably longer. By 1972, several other teachers had become involved in the development of science projects using students' interests and ideas. The concept was actively promoted by the Navajo Area Science Curriculum Committee and by Dr. David Warren, Director of the Research and Cultural Studies Section of the Bureau of Indian Affairs. An article published by Albert Snow in the October 1972 *Science Teacher* gave an additional opportunity to expose this concept to science educators. One point I should explain is that ethnoscience is not a National Science Foundation style of curriculum project at this time, with published materials and in-service and evaluation materials. The localized nature of the concept itself precludes this type of regimentation. It is an approach in which the local teacher and the community work together to produce curricular materials. Ethnoscience promotes the use of a specific set of cultural experiences in the teaching of concepts of science, but it also promotes the use of a rigorous science curriculum. The quality of the curriculum need not be compromised by the use of culture-related or culture-centered materials. We use a student's pre-science knowledge to show him that he already knows something about science and that science is not totally foreign to his experience.

The present status of the program is somewhat diffuse and not formal. There are science teachers in Alaska, New Mexico, and Arizona using an ethnoscience approach within their specific cultural setting. Likewise, there are elements of the concept that are used by various organizations, including the Navajo Health Authority, for recruiting Indian students into science-related careers. Our success or failure is not yet known. My own experiences with ethnoscience in the classroom have been successful, as have been those of at least two other teachers. It is a very time-consuming method for the teacher in that the curriculum frequently changes, since new materials must be developed as students pursue new problems. It is often slow work to develop the community input that is necessary to "certify" the validity of the materials used. However, community input is of critical importance to the development of a meaningful curriculum. Some of the specific methodologies of the ethnoscience approach to science instruction follow.

First, an essential tool is analogy. We use traditional and commonly known stories from the Navajo culture to explain relationships in nature. One good example of this is the tale of "Coyote and the Wildcat." In the story, Coyote plays a series of tricks on the Wildcat. In the end, however, Wildcat gets his revenge by feeding the Coyote some of his own intestines. This story is used in an environmental-science course to relate how man's abuse of his companion, nature, comes back to him. Likewise, other stories exist that are precise examples of succession, symbiosis, mutualism, and chemical reactions, to name but a few. This approach speeds up the teaching and allows more time for the more difficult course content. At the same time it also makes science a more humane subject without reducing the scientific content.

A second method used, as mentioned earlier, is that of involving students in research projects related to their own experiences. Students are all curious and will ask questions unless they are taught not to. These same students may become excellent laboratory assistants, and with the help of a teacher, they are usually quite willing to develop their research into instructional materials. With a little imagination, nearly any science topic can be taught with the use of local examples. One example of a student's research that has been used in our project was a study done on the impact of a fence on an open range. The data have been used in the classroom to study both the ecological and the economic aspects of fencing. This particular study received an award from the tribal chairmen.

A third method used sounds rather standard. We use numerous laboratory experiences for instruction, but wherever possible, we use local materials provided and prepared by the students. Again, the student research projects are a good source of information. A very successful laboratory manual has been written that uses a comparison of plants that grow in ditches and plants that grow in open fields.

These three methods should give some idea of the approach. Nearly any method of science instruction can be adapted. We have used simulations, student demonstrations, and slide–tape shows. It is important, however, to remember that certain methods may not be appropriate for certain students or cultures. For example, until recently, the Navajo language was not a written language, and therefore the Navajo child learned by listening, not reading. It follows, then, that the student's learning style was not initially writing-oriented but listening-oriented. This does not mean that a Navajo student should not be expected to learn from reading, but it does mean that it may be much more efficient to teach using tapes instead of books until the student's reading ability catches up with his learning ability.

I would like to suggest some generalizations using my experiences with ethnoscience for science education for minority students in other than a Navajo setting.

1. For a science-education program in a minority-population setting to be successful, there must be some community involvement. It might take many forms, but an interaction is necessary. In developing community involvement, one might ask the following questions in order to anticipate dealing with certain issues within the context of the community:

A. What are the culture-sensitive areas of the science curriculum?
B. What role will the science curriculum play in the community's change process?
C. What cultural experiences are appropriate to the teaching of contemporary science viewpoints?
D. What do students know about their traditional culture?
E. How can instructional materials be presented so that students understand both traditional and contemporary viewpoints and understand the setting of both?

2. Students themselves are the best ambassadors for the science curriculum and for science careers. A young college student in a premedical program or a preengineering program or a technician's training program can relate to high-school students. Likewise a student working on a student research project that he in turn presents to a class in some form of a lesson via videotape or a slide–tape show communicates to students their own potential in science.

3. We must remove the idea that the only careers in science worth promoting are those in the laboratory and those requiring advanced degrees and that being in a science career means being a physicist or a biologist or some similar type of highly trained, highly specialized

person. The impact of this approach is to scare off students who eventually would successfully complete advanced degrees and might enter the laboratory. It might even be true that the science teacher, not the scientist, is the most important link to emphasize at the present. It is the teacher who promotes science with the public and academically nurtures the future scientist.

4. As suggested above, science educators, in my viewpoint, are the link in promoting science and science careers to minority students. To perform this role effectively, the teacher must have certain characteristics. These are:

A. An awareness of the social organization of the community and the students.
B. A knowledge of traumatic events that may have a major impact upon a student's response to new ideas in the science classroom.
C. A knowledge of the taboos and the etiquette of the community and the students and a working knowledge of the stories and the traditions of the community.
D. A knowledge of the environmental and technological characteristics of the community and the location of technological resources for the classroom.
E. A knowledge of what people and what organizations have the greatest interface with the dominant culture, both inside and surrounding the community. These might include planning boards, museums, health authorities, etc.

What I am suggesting in these points is a humanistic approach to teaching that goes beyond content and method and includes the experiences of students as central, not peripheral, to the science curriculum.

In summary, I suggest the following:

1. The development of programs to improve minority science education and to encourage minority students to enter science careers should be undertaken with a larger context in mind. Using shotgun approaches or a very limited view may undermine the very strengths that many cultural minorities see as essential to their survival and that I see as important to the health of the American culture.

2. Most students already know a good deal about either the content or the concepts of science. However, we need to realign this knowledge so that students realize that they know it and so that they see it within

the context of science. As a result, these students will not be bored with science classes or be afraid of science careers.

3. Science educators should incorporate humane methods of instruction into their curricula and develop more region- or culture-specific programs. All scientists do not develop their activities by identical means; therefore, all teachers need not teach in the same monolithic pattern: alternate considerations should be given to learning styles and content.

4. And last, communities or at least some members of local communities should be involved in the science-education process. It makes little difference what the involvement is initially, just as long as there is involvement. In the case of ethnoscience, our community members reviewed course material, compiled resource material, and served as instructors.

In conclusion, I would like to make a comment on the ethnoscience concept. The ideas suggested here may be relevant not only for minority students but also for the general American public. For some time now, articles in *Science* have bemoaned the reduced support of science and science education by the public. The ethnoscience approach to science instruction might be especially useful in removing the separatist feelings of the public and in developing in students a usefulness for both what they have learned and what they might contribute to the scientific enterprise in their own community.

Science is not a substance that can be handed out. It is rather a process to which certain kinds of jobs contribute and for which a whole series of events and people are needed. Culture-specific science education can do much to set the stage for the process and to encourage cultural minority participation in the training of science professionals.

32

The Utility of a Piagetian Approach in Developing Precollegiate Science Programs for Minorities

SIDNEY A. McNAIRY, JR., and BOBBIE N. McNAIRY

From the moment a child is born, a life-long mental process is initiated in which there are many mental syntheses, combinations, resyntheses, restructurings, and reorganizations. Even at the outset, the child's reasoning process occurs through an ordered sequence of events. This sequence allows the child to interpret the social and physical events that he or she encounters for the remainder of his or her life. Thus, even at this early stage of mental development, theories, although they may appear elementary to the adult mind, are formulated and serve as the initial basis for cognitive development.

These ideas are by no means novel; they belong to the discipline of structural-developmental psychology. This discipline was placed on a firm psychological basis by the Swiss psychologist and philosopher Jean Piaget and was widely proclaimed in this country 15 years ago (Jennings and Piaget, 1967). Piaget is a proponent of the school of thought that believes that intellectual development is very similar to enzyme induction in biological systems. Just as enzymes are produced by the biological system to deal effectively with the catabolic and anabolic processes of metabolism, the mind is constantly involved in a similar process as it catalogs various stimuli into cognitive responses. Based on his studies of mathematical and logical models, Piaget pro-

SIDNEY A. McNAIRY, JR. • Health Scientist Administrator, Minority Biomedical Support Program, Division of Research Resources, National Institutes of Health, Washington, D.C. BOBBIE N. McNAIRY • Instructor of Mathematics, Prescott Junior High School, Baton Rouge, Louisiana.

posed that intellectual development occurs in the following stages:

1. Sensorimotor intelligence, 0–2 years.
2. Preoperational thought, 2–7 years.
3. Concrete operations, 7–11 years.
4. Formal operations, 11–15 years.

Each of these stages of development is characterized by specific capabilities and inabilities. Since we are concerned with the precollegiate student in this symposium, let us focus in on the last two stages of development, since these are the principle stages of development for most high-school students.

During the period of "concrete operations," the individual student lacks the mental sophistication to apply working knowledge to hypothetical situations. Hence, the student lacks the ability to utilize a given set of facts in making predictions about the future. On the other hand, during the period of "formal operations" and thereafter, a student has the mental sophistication to deal with the past, the present, and the future, to assimilate the knowledge at hand, and to rearrange it into working hypotheses. In contrast to the "concrete level" of development, at which time a student can only serialize, extend, subdivide, or combine bits of knowledge in new relationships, the student who has advanced to the "formal level" of thinking possesses abstract reasoning skills that can be used in the systematic manipulation of variables. This student is thus able to utilize acquired knowledge in solving problems.

McKinnon (1971) found 50% of the college freshmen he tested to be below the "formal level" of thinking on at least one Piagetian task. These same observations were supported by Griffith's comparative studies with college students at Rutgers and Essex Community College. These studies showed little difference in the mental development of Rutgers University freshmen and inner-city college students. At least half of these students were at the "concrete level" of thinking. Just as importantly, Milton Schwebel (1972), in his report to the U.S. Office of Education entitled "Logical Thinking in College Students," indicated that the majority of college students are below the formal level of thinking. Thus, it is no wonder that George Koloidy (1975) reported that 7 of the 25 college freshmen subjects in his studies failed to complete the second semester of physics (four changed their majors and three did not reregister).

The majority of college students do not possess the mental capabilities needed for problem solving associated with science courses. Accordingly, many students fail to pursue a career in science because of this deficiency at the critical stage of mental development. Others who started to college with an initial desire to pursue a career in science often end up in nonscience majors. Many college professors fail to recognize that even though the educational experience must be sufficiently novel to create mental disequilibriums, it must at the same time have some relevance to what the student knows and to the student's stage of mental development. Classroom activities and other exercises should create a systematic union between the student's mind and the concepts that are being taught. Hence, it is often necessary to develop educational materials that have a broad range of mental stimulations that can reach a group of students whose level of mental preparation varies over a wide range of cognition. Thus, in many instances, the educational experience must, out of necessity, be oriented more toward the individual than toward the group. More often than not, the traditional lecture approach to teaching can be self-defeating unless it is complimented by exercises that allow students to interact verbally with individuals who are at similar stages of mental development. These kinds of student interactions serve as excellent sources of "teaching assistants" who are more closely associated with individual students.

Based on the above observations, there is without question a very urgent need to elevate the level of cognition of the beginning college freshman, especially if there is a serious commitment to increase the number of minorities in the sciences.

The data in Table 1 show the dearth of minorities in science and engineering. The table shows that the total minority science population was 53,500 of the 1.3 million scientists and engineers in the United States labor force in 1972. Of this total number of minority scientists, there were approximately 29,000 engineers and 25,000 scientists. Of all minority scientists, 60% were of Oriental descent, 30% were black, and the rest were members of other nonwhite groups.

The situation is even more grave when one is made aware of the fact that there are approximately 2,000 black Ph.D.'s out of a total Ph.D. population of 215,000. There are approximately 800–1,000 black Ph.D.'s in the natural sciences. Just as startling is the fact that there are only 6,000 black physicians out of a total of 350,000 physicians. Chicanos,

Table 1. Minorities in the Sciences: Total Population, 1970, and Scientists and Engineers, 1972, by Racial Group[a]

Total population, scientists and engineers	Total	White	Minorities: Total	Minorities: Black	Minorities: Oriental
(1970)			Millions		
Total population	203.2	177.7	25.5	22.6	1.4
(1972)			Thousands		
Total scientists and engineers	1,336.4	1,283.0	53.5	16.2	32.1
Scientists	496.1	471.3	24.8	9.7	12.8
Engineers	840.3	811.7	28.6	6.5	9.3

[a]Source: National Science Foundation (1975).

American Indians, Hawaiians, and Puerto Ricans also are inadequately represented in all categories. The educational plight of the Mexican-American is extremely grave. According to Fernando C. deBaca, Special Assistant to President Ford, "Mexican-Americans are the most education-deprived group in the Nation." In a report to the President in 1975, he made the observation that it would take "a 330 percent increase in Mexican-American College and University enrollment for equitable representation of the Mexican-American."

Chicanos and other minorities as a group score significantly lower than whites on the standardized tests that are used for admission to college. According to a task force from the University of California, a primary reason for the poor performance by Chicanos is the culturally biased standardized tests; the same probably holds true for other minorities. These tests are designed to evaluate the intellectual achievements of English-speaking, middle-to-upper-class white students. Even after entering college, Chicano students have difficulty with English, and according to 1973 figures from the College Entrance Examination Board, only one out of four Chicanos who enter college actually graduates; in the case of whites who are admitted to college, one out of two graduates. Census Bureau figures for 1970 show that only 58 out of 1,000 Chicanos aged 22 or older have had four or more years of higher education.

It is therefore not amazing but quite alarming that minorities represent less than 4% of all the scientists and engineers in this country. Hence, the task of increasing the number of minority scientists and engineers is both a mammoth and an urgent one. It is absolutely essential that the pool of precollegiate minorities interested in careers in science and engineering be expanded. In order to have any significant impact, programs that are undertaken should be national in scope. Piagetian methodologies should be used to prepare the precollegiate student to pursue a career in science and engineering.

THE PRECOLLEGIATE SCIENCE PARTNERSHIP (A PROPOSED PROGRAM)

Just as the medical colleges of Meharry and Howard University have provided the major impetus for the education of black physicians and dentists (they have been responsible for training 80–85% of all the black physicians currently practicing medicine), the historically black college has also played a major role in developing black scientists and engineers. These black colleges have been the major sources of the black undergraduate students who have successfully pursued the doctorate degree in major universities. The expertise that the faculty at these black colleges have been able to acquire through their many years of on-the-job training with minority students and their resulting sensitivity to the various motivational needs of minority students will be invaluable assets to precollegiate programs for minority students. The program that is proposed here could also be implemented at majority institutions wherever there are significant numbers of minority students and a commitment to assist in eliminating the dearth of minority scientists and engineers in this country. Initiating the proposed program at committed majority institutions will make available additional avenues through which black, as well as Chicano, Native American, Puerto Rican, Oriental and other minorities, can enter the science and engineering professions.

The proposed program consists of the following student activities: (1) attending daily nontraditional lectures that last for two hours, accompanied by guided discovery exercises, and (2) working in a research laboratory daily for at least four hours. The program should take place over a nine-week period during the summer and should allow high-school students who have completed their sophomore year to start their initial participation in the program. If the program is to have significant impact on the cognitive development of precollegiate students, each student should be encouraged to participate in the program the summers following the junior and senior years of high school.

Lectures. The participating high-school students attend daily lectures that teach them to take a series of observations, determine regularities, and then formulate hypotheses. This kind of approach serves to provide the students with the necessary foundation for problem solving so vital in research investigations. Hence, the lectures should allow the students to extend their thinking capabilities from the "concrete operational level" to the "formal level." Thus, students who can only serialize, subdivide, or combine pieces of knowledge in new relationships will develop into students who can apply abstract-reasoning skills while using concrete knowledge in manipulating variables in a systematic way. It is expected that the topics used in the course will serve to facilitate the development of the students' analytical thinking on the formal level. For other students, it is expected that there will be a rebirth of formal thinking abilities. Thus, the lectures should involve an in-depth discussion of concepts, each of which should be followed by the assignment of relevant chemistry, biology, or physics problems. These problems should require the use of concepts from the lecture. Some examples of concepts that lend themselves well to the development of guided discovery exercises are: (1) oxidation/reduction; (2) indicators and *p*H; (3) Beer's law; (4) molecular weights and ultracentrifugation; (5) energy transformations in the human body; (6) electricity and magnetism; (7) discgel electrophoresis; and (8) ion-exchange chromatography. The details of this approach may be found in the approach that the faculty at Southern University has used in their chemistry courses for nonscience majors (see Moore, McNairy, and Bursh, 1974).

Research. The students' activities in the laboratory should be divided into two phases. During the first three weeks, the high-school participants should work along with an upperclass undergraduate student in order to learn the specific technique necessary for conducting the "mini" research project assigned each high-school student during the remaining six-week period. The first phase of the high-school student's experience in the laboratory, as well as the second phase, should be under the constant supervision of a research director in the laboratory. Ideally, the high-school student's undergraduate research partner should be a college sophomore or junior so that there will be continuity in the high-school student's research experience when he returns to the program the following summer.

An excellent national program that can serve as an appropriate example for implementing the proposed program is the Minority Biomedical Support (MBS) program at the National Institutes of Health. The MBS program has a history of aiding the development of minority scientists and has an emphasis on undergraduate research. The range of institutions now involved in the MBS program covers a broad spectrum. Although the majority of colleges involved are black colleges, the program also exists at nonblack institutions where there are a significant number of minority students and a commitment to assist in increasing the number of minority scientists. This program, which emphasizes undergraduate research, has had phenomenal success already. In 1975, for example, of the 293 MBS graduates, over 75% went to graduate or professional school. As a direct result of the existence of 410 MBS research projects at 78 different institutions, 75 scientific articles were published in reference journals, and 358 scientific papers were presented at various scientific meetings in the United States and several foreign countries.

REFERENCES

Jennings, F., and Piaget, J. Notes on learning. *Saturday Review,* May 20, 1967, pp. 81–83.

Koloidy, G. The cognitive development of high school and college science students. *Journal of College Science Teaching,* 1975, *5,* 20.

McKinnon, J. W. Earth science, density and the college freshman. *Journal of Geological Education,* 1971, *19,* 218.

Moore, W. E., McNairy, S. A., and Bursh, T. P. Interacting inquiry. *Journal of College Science Teaching,* 1974, *3,* 271.

National Science Foundation. Science resource studies highlights: Racial minorities in the scientist and engineer population. September 19, 1975.

Schwebel, M. Logical thinking in college students. Washington, D.C.: U.S. Office of Education, 1972.

33

Science Education for Minorities: A Bibliography: A Preliminary Report

SHIRLEY MAHALEY MALCOM

Science and applied science are deeply embedded in the history of the development of this country. From Ben Franklin's kite experiment to artificial kidneys to men on the moon, much of our pioneering spirit has been directed toward exploring the frontiers of science and technology.

The history of minority participation and involvement in science is shorter and somewhat more obscured. This is not to say that there were no "early American" minority scientists, engineers, and physicians. Individuals belonging to the various racial–ethnic minority groups have contributed much to these fields, but entrance into the so-called mainstream of scientific activity in the United States has been much more difficult or has gone unnoted. It took 100 years from its founding for America to produce its first black Ph.D. in science. A hundred years later, blacks and other minorities are yet to enter science careers in significant numbers. Though many of us are in science today partly as a spin-off of the *Sputnik*-induced science explosion, it is only recently that any concerted efforts to bring minorities into science have been initiated.

During the last seven years, many programs have been developed to bring minority Americans into the mainstream of biomedical science. Because of the localized nature of many of these efforts, very little communication exists between programs, in both their development and their implementation (programs developed by the National Institutes of Health and the National Science Foundation are exceptions to

SHIRLEY MAHALEY MALCOM • Staff Associate, Office of Opportunities in Science, American Association for the Advancement of Science, Washington, D.C.

this statement). This lack of communication has reduced the possibility of sharing experiences for those involved in such efforts.

As an initial attempt to solve the communication problem, the Office of Opportunities in Science and the Office of Science Education of the American Association for the Advancement of Science proposed the development of *Science Education for Minorities: A Bibliography.* This project is supported by the National Science Foundation. Such a listing could provide a directory for those interested in identifying experiences in science education for minorities. In addition, it could prove to be a first step in a national move to set priorities and suggest relevant future strategies for involving minorities in science.

Although information exists that indicates that most minority students are lost from the pool of potential scientists, engineers, and physicians prior to entrance into college, only a relatively small proportion of the minority science-education efforts have been directed toward the precollege level. Of the more than 300 programs already identified and cataloged, fewer than 60 were directed toward the elementary- and secondary-school student.

Following is a review of some of these programs.

SINGLE-EXPERIENCE PROJECTS

Many organizations have attempted to provide an experience in science or an exposure to science careers for minority students. For example, the Latin American Women of the Museum of Science in Miami have established a program that provides museum and planetarium experience to the children of migrant workers. In a similar manner, the Science and Mathematics Career Day sponsored by the University of Massachusetts at Boston attempted to provide science-career information via laboratory sessions, workshops with films, and other resource materials. A major feature of these programs was the involvement of minority professionals as role models. The disadvantage of such programs is the lack of follow-up or reinforcement of the experiences. The advantage of this type of program is the large number of students that can be involved for a fairly small investment in time and money. Other organizations, such as zoological parks, botanical gardens, and hospitals, could adapt such one-day experiences to their particular situations. The learning experience, coupled with the involvement of minority professionals as role models, could have a significant impact in terms of supplementing the regular school and counseling experience.

TEACHER-TRAINING PROGRAMS

Another way of influencing a large number of minority students is through their teachers. Inner City Teachers of Science (ICTOS) of Brown University seeks to train secondary-school science teachers of urban schools by means of specifically developed science courses, a special field-based course on teaching in the inner city, student teaching, and a course in improvising equipment for teaching science in schools with low budgets for science and related courses. This program attempts to inculcate in its trainees a sensitivity to the special problems of inner-city pupils and schools. In a similar manner, the Urban Science Intern Teaching Project in Los Angeles seeks to identify, educate, and place individuals who have an interest in and an aptitude for teaching science to the educationally uninvolved students in inner-city schools.

Many projects in science-curriculum implementation have involved the in-service training of science teachers from school districts with large minority populations. These teachers are trained in the use of the various new curricula, such as the Science Curriculum Improvement Study (SCIS) and Science—A Process Approach (SAPA). These curricula were developed initially to improve the overall quality of the science instruction that students receive at the precollegiate level. They involve "hands-on" discovery methodology; many require no reading by the student and are thus extremely valuable in stimulating an early interest in science in minority students who lack verbal skills. Another project has coupled the approach of science-curriculum implementation with bilingual training and has provided Spanish translation for activities, vocabulary, etc. The San Antonio School District implemented a bilingual education program that sought to incorporate language considerations, the linguistic peculiarities of Spanish relative to science, and the learner characteristics of the Mexican-American student in the teaching of science.

COUNSELOR-TRAINING PROGRAMS

Some programs have been directed toward counselors in minority schools—exposing them to information on science careers, opportunities, and available financial aid. A program undertaken in Los Angeles schools involved the production and use of short filmstrips for in-service counselor training and workshops (sensitivity sessions of a sort that dealt with career choices in science vis-à-vis racial and sexual stereotyping).

CURRICULUM- AND COURSE-DEVELOPMENT PROGRAMS

Project City Science, based in the New York public schools, is an example of an innovative curriculum program directed at improving the quality of the science education that minority students receive.

A low-cost elementary science curriculum was developed at the Lawrence Livermore Laboratory of the University of California for use in the Oakland and other East Bay schools. The Lawrence Livermore Laboratory Elementary Science Study of Nature (LESSON) was developed because of a need for inexpensive science education in the inner-city schools.

Engineering, A Piece of the Action, developed by the DuPont Company and the Del Mod System, is a series of lessons for use in junior high school to motivate students toward engineering and/or related technology as career choices.

SCIENTISTS AS TEACHERS

Innovative projects have involved the use of industrial scientists in public-school teaching situations. For example, the Xerox Science Consultant Program sends scientists into the Rochester inner-city schools, where they conduct scientific experiments for fourth-grade through sixth-grade students. This program also involves sending scientists into schools for physically handicapped students.

STUDENT-ORIENTED PROGRAMS

The most commonly encountered approach is to deal directly with the student, influencing the total educational experience, especially the science component of the experience and especially on the secondary-school level. Such programs range from a mathematics–science high school in an inner-city setting to summer, after-school, and/or weekend programs designed to enrich and supplement the regular experience in school. These programs usually involve several or all of the following components: laboratory experience; tutorials; seminars by professionals (most use minority professionals extensively); minicourses; field trips to industries, hospitals, etc.; counseling; and career education. Such programs are usually higher in cost per student than those mentioned previously. Some of the most comprehensive programs are developmental in nature in that they are designed for the entire high-school career. The disadvantage of higher financial output per student usually means that fewer students can be involved in such programs, but careful initial screening and comprehensive, longitudinal followup seem to yield a greater probability of success in the long run (as measured by the percentage of students entering and succeeding in college science curricula). Most of these comprehensive, longitudinal programs are aimed at channeling students into the engineering fields; some others are aimed at increasing the number of students entering the health professions. (For more information, see *An Inventory of Programs in Science for Minority Students, 1960–1975,* American Association for the Advancement of Science, Publication No. 76-R-10.)

Comparable "total-experience" programs on the elementary level are the numerous environmental education programs that have been undertaken in various parts of the country. Most involve summer-camp experiences that are supplemented by minicourses in environmental sciences.

MEDIA

Books have been written, good films produced, and short commercial messages developed to motivate minority students toward careers in science, engineering, and the health fields, but these are few in number.

Many educators have proposed the effective use of commercial television programming as a teaching tool that could reach hundreds of thousands of young viewers, including preschool-aged boys and girls. A program like "Infinity Factory" or "Sesame Street" that has a strong science component and features minority scientists in a format likely to attract children could go a long way toward shaping the attitudes of minority youth about science and scientists.

The inventory of programs in science for minority students is scheduled for completion in the early summer of 1976.* Its usefulness as a directory is obvious. Its greater potential can be realized only from an evaluation of the programs included; in this way, it can be a possible first step in a national assessment of science education for minorities. If we are to increase the participation of minorities in science, we must increase the numbers of young people who are interested in careers in science. We must be concerned with the quality of elementary and secondary science education. We must be concerned that minority students are not turned off from science before they have the opportunity to consider a career choice in science. We must be concerned that minority students have every opportunity to consider a variety of career options. We need a comprehensive and systematic approach to science education, especially for minority students.

*Available from the Office of Opportunities in Science, American Association for the Advancement of Science, 1776 Massachusetts Avenue, NW., Washington, D.C. 20036.

34

Summary

SHIRLEY MAHALEY MALCOM, WAYNE FORTUNATO-SCHWANDT, FRANKLIN D. HAMILTON, and VIJAYA L. MELNICK

The papers presented in this volume represent the views expressed by concerned educators and students, administrators and scientists, and representatives from public and private sectors on the question of minorities in science. The common thread that runs through the papers is an effort to call attention to the current, severe underrepresentation of minorities and the disadvantaged in the biomedical professions, and the need to correct this inequity.

In American society education is a key factor for the improvement and enhancement of opportunities for social advancement. Prejudice and discrimination, based on the scarred history of racial strife in this nation, have perpetuated an unequal educational system coupled with unequal opportunities for entry and advancement in the professions. Even though laws exist that can do away with much of what is wrong and unjust, the enforcement of these laws has been unduly slow. Such laws have often been challenged with narrow, local, and petty legal-technical objections that have caused near paralysis to just and needed progress. The burden of understaffed and underequipped schools has fallen largely on the minority communities who, in general, can boast of little political or economic clout.

Programs are discussed that are aimed at supplementing the academic-scientific skills often found deficient in students coming from small underfinanced schools. The high degree of success experienced

SHIRLEY MAHALEY MALCOM and WAYNE FORTUNATO-SCHWANDT • Office of Opportunities in Science, American Association for the Advancement of Science, Washington, D.C. FRANKLIN D. HAMILTON • Associate Professor, The University of Tennessee–Oak Ridge Graduate School of Biomedical Sciences, Oak Ridge National Laboratory, Oak Ridge, Tennessee. VIJAYA L. MELNICK • Associate Professor of Biology, Federal City College, Washington, D.C.

by such special programs, measured in terms of graduate student placement, speaks not only of the caliber of such programs but also of the potential and the high degree of motivation demonstrated by these students. Efforts to increase the number of minorities in science and health professional fields can be separated into two broad categories. The first category contains activities designed to make the elementary and the secondary school systems more cognizant of and responsive to the needs of minority students in helping them plan career opportunities. Papers by Watson, Jay, and Warren address the need for grade schools to offer appropriate curricular material and to equip teachers with skills that would give all students better preparation for college study. Changes in the grade-school system would provide a long-range solution to the problem and would result in a significant increase in the number of minority students interested in and prepared for college science training, thereby increasing substantially the pool of potential science and health professionals.

Another realm of activity that will have a long-range impact on the movement of minorities into science areas has to do with standardized testing. Tests are used nationwide throughout student development to determine student potential for success. Test scores are significant criteria for admission to college and graduate and professional schools. Martinez, Maestas, and Warren speak of the traditionally poor performance on standardized tests by students with bilingual backgrounds and poor verbal abilities. In many cases, poor performances result not from a lack of knowledge on the subject matter but from a lack of familiarity with the testing mechanism. The papers of Angel and Nieves indicate that admission committees should use test scores as only one determinant for measuring a student's capability to perform on Graduate Record Examinations. Nieves points out that the Educational Testing Service has started to compile test data on minority students using the GRE in an effort to determine if standardized tests can be as effective a measure of minority students as it is of majority students. Because of their concern of the standardized test as an adequate predictor, the American Association for Medical Colleges has undertaken a complete revision of the Medical College Admission Test. In his paper Angel reports on the new exam, the Medical College Assessment Program (MCAAP), and indicates how it is designed to insure inclusion of only those factors judged to be relevant to becoming a good physician.

Papers in Section II address the problems and difficulties minority students encounter at majority institutions. Quite frequently these institutions do not provide an adequate supply of academic support mechanisms, i.e., counseling and tutoring. Many of these institutions have an insufficient number of minority faculty members who can serve as role models and liaison with other faculty members who, because of cultural differences or biases, may not perceive the particular needs of minority students. Often they lack also a critical group of minority students essential for maintaining a healthy social environment. These factors tend to increase the normal anxieties and stresses that students experience in difficult academic programs.

Papers in Section IV address problems that are encountered in enforcement of equal-employment statutes. Affirmative action in employment has been hailed by some as "too little" and by others as "too much." However, the "old buddy" network, which does not include minorities, continued to function in academia and serves as a barrier to equal-employment opportunities. Roybal speaks of a need for more aggressive enforcement by Federal Agencies and forthright implementation by academic institutions of civil rights legislation. Goodwin cites various examples of "games" that are played by institutional officials to evade affirmative-action measures. Gordon offers several recommendations that would lead to effective enforcement of affirmative-action policies by governmental agencies.

A second category of actions which will bring a more immediate solution to the inadequate representation of minorities in science can be obtained via two approaches. The first involves special-training programs for minority students. Papers describing various types of programs are found in the last two chapters of this volume. These programs involve high school, undergraduate, or postbaccalaureate students and are designed to enhance students' awareness of and preparation for careers in science. Programs of these types are funded by many organizations, both governmental and private. Other papers describe the funding mechanisms and programs in the National Institutes of Health, the National Institute of Mental Health, and the National Science Foundation. It is essential to maintain funding for special programs for the immediate future to ensure a pool of minority students who will qualify for and matriculate in science- and health-training programs.

A second approach to increasing minority participation in science presented in papers by Evans, Melnick, Carew, and Roybal is that of effecting change at the policy-making level of governmental institutions. Evans states that members of minority groups must attain access to positions with the power to "define." If the policies and priorities of agencies which control the resources for research and training are to show a genuine interest in bringing about a full participation of minorities in biomedical professions, then such policies should reflect the need for making sure that there is a good representation of minorities both within the agencies and on their advisory boards.

Both approaches discussed above are necessary to bring about an increase of minorities in science. Both special programs and policies are needed, and they require the attention to those individuals interested in the ways in which our scientifically and technologically based society serves minority peoples.

Should there be concern that minorities are underrepresented in science and biomedicine? This question is answered in several different ways by the authors. Many point to the need to redress past discrimination and the provision of access to research careers. Others point to the fact that the minority community is largely served by minority health professionals and that the underrepresentation of minorities in these professions results in inferior health professionals and inferior health care for minority communities. There is an overwhelming need to have trained minority professionals who can participate in decision-making processes in scientific–technological–health areas that will concern or affect minority communities. The presence of jobs and higher salaries in these areas can contribute to reducing the gap between the incomes of minority and majority families. In addition, they note that minorities are not adequately represented on faculties of major colleges and universities in the United States. As Goodwin states, "the purpose of affirmative action is to raise standards, not to lower or maintain them. Standards are raised not only by an honest, broad, innovative search that includes all the competent minority and women candidates available but by the establishment of cultural pluralism in our faculties. . . . Diversity and quality are compatible and essential."

The need to produce more minority physicians, dentists, and other health professionals is clearly documented. That shortages of such individuals exist in minority communities is evident from the data. However, the production of more minority Ph.D.'s in the sciences has been questioned by those who see an oversupply of science doctorates in the nation as a whole. Jay addresses this issue by pointing to the fact that the oversupply may be valid for white Americans but not for black Americans. And this is also true for Latino and American Indian groups. Martinez cites the dismal figures for "Spanish-surnamed Americans" in science and engineering. There are fewer than fifty American Indian science doctorates in this country. The record speaks for itself, there is no danger of oversupply of Ph.D. scientists and engineers who are American Indian, black, Mexican-American, or Puerto Rican.

Some of the specific suggestions that will produce a change in the patterns of representation of minorities in science and biomedicine can be listed as follows:

1. Initiate a major curriculum change and teacher-training effort at the elementary school level.
2. Provide bilingual education at the elementary and secondary school levels where needed.
3. Improve the counseling system to alert minority students of career potentials and capabilities and of job opportunities available in the biomedical and health-related professions.
4. Provide teachers with knowledge about different cultural values and styles.
5. Emphasize the development of good mathematical, reading, and communication skills early in a student's training.
6. Develop laboratory-based science curriculum to increase student motivation.
7. Provide incentives and necessary role models for students.
8. Expand financial support for science research and training for minority students at both minority and majority institutions.
9. Provide needed tutorial and counseling services for minority students at majority institutions.
10. Increase the numbers of minorities serving as faculty members at majority institutions.
11. Encourage and expand special training programs for minority students who are interested in careers in science and/or biomedicine.
12. Expand science-related, work–study, and undergraduate research experiences. (Programs of this type should be developed for both the physical sciences and engineering.)

13. Increase the recruitment of minority students for medical, graduate, and professional schools.
14. Give adequate attention to a student's potential in the assessment of his or her qualifications for graduate or professional school.
15. Keep minority communities informed about scientific and health issues that affect them.
16. Vigorously enforce equal-employment opportunity and affirmative-action legislation.
17. Make advisory committees and policy-making bodies more representative of minorities.

Name Index

Subject Index